Endorsements

"Dr. Hamden's fresh, funny take on our wildly strange universe will leave you laughing...while also expanding your understanding of our place in the cosmos."

—**Michelle Nelson**, executive producer of *Mission Unstoppable* on CBS

"*Weird Universe* is a brilliant, raucous delight. With her gift for storytelling, Dr. Hamden reveals the quirks and wonders of our universe in a way that's simultaneously laugh-out-loud funny and deeply informative. This book is a must-read for anyone, science lover or not, who wants to explore just how strange space is. Prepare to laugh, learn, and look at the stars a little differently."

—**Dr. Sarafina El-Badry Nance**, author of *Starstruck: A Memoir of Astrophysics and Finding Light in the Dark*

WEIRD
UNIVERSE

WEIRD
UNIVERSE

Everything We Don't Know About

Space (and Why It's Important)

Dr. Erika Hamden

MIAMI

Cover Design: Elina Diaz
Cover Photo/Illustration: stock.adobe.com/angel_nt
Author Photo: Bret Hartman / TED
Layout & Design: Elina Diaz

For permission requests, please contact the publisher at:
Mango Publishing Group
5966 South Dixie Highway, Suite 300
Miami, FL 33143
info@mango.bz

For special orders, quantity sales, course adoptions and corporate sales, please email the publisher at sales@mango.bz. For trade and wholesale sales, please contact Ingram Publisher Services at customer.service@ingramcontent.com or +1.800.509.4887.

Weird Universe: Everything We Don't Know About Space (and Why It's Important)

Library of Congress Cataloging-in-Publication number: 2025931824
ISBN: (print) 978-1-68481-728-3, (ebook) 978-1-68481-729-0
BISAC category code: SCI015000 SCIENCE / Space Science / Cosmology

To the stars, the moon, and the night,
illuminating the way for all who look up.

Table of Contents

Introduction..11

Chapter 1
Spacetime, the Counterintuitive Fabric of the Universe..................16

Chapter 2
Black Holes: A Mathematical Novelty That Turned
Out to Be Real. Also, Not a Vacuum Cleaner..................................25

Chapter 3
Space Is Actually Quite Large (Larger Than You Think)..................37

Chapter 4
Stars, the Universe's Response to Gravity.......................................47

Chapter 5
The Biggest, Fastest, Smallest, Weirdest Stars................................60

Chapter 6
Nuclear Spaghetti and the Deaths of Stars.....................................74

Chapter 7
We Used to Think This Was All There Was.......................................83

Chapter 8
Up Until 1960, We Thought Venus Was a Tropical Paradise............98

Chapter 9
Planets Are Surprisingly Common...109

Chapter 10
Most Solar Systems Are Nothing like Our Own..............................118

Chapter 11
Every Planet We Reach Is Dead..127

Chapter 12
Quasars, Rings, and Galaxies at the Beginning of Time................136

Chapter 13
Speaking of the Beginning of Time... 150

Chapter 14
The (Tragic and Yet Possibly Quite Boring)
Heat Death of the Universe 162

Chapter 15
The Dark Sector: The Weirdest Items on a Large but
Incomplete List of All the Things We Don't Know 172

Chapter 16
We Are the Universe Observing Itself 181

Conclusion
The Earth Isn't at the Center of Anything 188

About the Author 190

Introduction

How weird is the universe?

To give you the short answer, the universe is pretty weird.
Congrats on getting the main takeaway of this book.

For a long time, we didn't know how weird the universe was. As
long as there have been humans, they have looked up at the night
sky and wondered about what they saw there—the pinpoints of
light and the carpet of stars. Early humans made up myths and
stories about why the stars were in the sky and every culture
developed a cosmology to explain why we were here and the
presence of the Sun, the Moon, and planets.

As humans first started to slowly figure out our place in the
universe, trying to determine the structure of the solar system
and the universe around us, the assumption of normalcy was
paramount. We've always taken our experiences here on Earth
and extrapolated them outward into space. In some sense, this is
because we are flawed humans trying to understand something
nonhuman. Our brains like to find patterns and so we spend a lot
of time trying to fit the universe into the patterns we see on Earth.
We didn't know any better.

An example: an early theory of the Sun's composition was that it had the same composition as the Earth. All planets had the same composition and somehow, they had spun off from the Sun at some point like little droplets of water off a rapidly spinning sphere. This turned out to be wrong, both on the Sun's composition and on the process of how planets are made. The Sun's composition (described in a later chapter) is so unlike Earth's composition that we found a new element there (helium, named after Helios, the Greek god of the Sun) before we isolated it on Earth. We now know stars are fundamentally different than planets and far stranger.

Another example: Sound waves travel through air here on the surface of Earth. Waves on the ocean travel through water. To a person living on Earth, it makes perfect sense to assume that all waves travel through a medium. Thus, early physicists assumed that light waves moving through space must travel through a medium evenly distributed throughout space. They dubbed this medium "luminiferous aether," or "the ether," and said it should fill all of space if space is to make any sense and behave like things on Earth. Scientists conducted numerous experiments to detect the presence of the ether and did not find any hint of it. They revised and updated theories to account for these non-detections, but the new theories became increasingly hard to believe. After existing as a concept for hundreds of years, "the ether" was retired, and a much weirder conclusion was reached: Light doesn't need a medium to travel through and light is nothing like sound waves or ocean waves. Light only ever travels at one speed. Light is both a wave and a particle. And light has no mass but only energy. And photons, the particle form of light, don't experience time as we do. And light can be converted into mass when needed or can be created out of mass when the conditions are right. Light, a ubiquitous part of our experience of life on Earth, is so unbelievably weird that it took hundreds of years of study to figure out even some of its most basic properties. We still don't understand it.

This story is repeated countless times in the history of science and our understanding of the universe. The first assumption is objects in space behave like objects on Earth. Or our experiences here on Earth, traveling slowly through spacetime, are universal and represent the whole of reality. At every opportunity, the universe has shown us that those assumptions are wrong and, worse, tragically boring. Compare the mundane concept of air in space (just like on Earth) versus the mind-bending reality of the vacuum of space, the weirdness of light, and spacetime. Compare the Earth as *the* fixed, unmoving center of the universe to the reality of our actual universe, which has no center yet was still created from a single point.

Even the vastness of space is weird. A big argument in astronomy at the beginning of the twentieth century was whether the universe was larger than our galaxy or if our Milky Way galaxy contained all that there was. Looking back on this, how small and miserable to assume that the one galaxy you happen to live in contains all that there is.

Nothing about our experiences here on Earth can prepare us for how weird the universe is. Our desire to make space behave like life on Earth—constrained, small, easy to understand, human-centered—has been thwarted at every opportunity by the universe itself. Maybe this is because of the self-absorbed nature of humans: we all live in a world with us at the center. The most important people are all humans; the most important planet is Earth. Of course, we take a human-centered view of things when we start to explore. We are the fulcrum around which all things turn.

But happily, the universe itself is vast and indifferent to our small human-focused desires. Instead, the universe has given us a seemingly endless bounty of weirdness to discover—the nature of energy, the composition of stars, the fabric of spacetime, the beginning and end of time, and the quantum mechanical nature of all atoms. As we learn more about the universe around us, we find that it's even more wacky than we could have imagined. So weird, in fact, that it has taken scientists (that's us) time to get comfortable with it. People hated quantum mechanics when it was first developed. The person who coined the term "The Big Bang" for the start of the universe did so with scorn—he thought the idea was so preposterous. It took decades for scientists to be comfortable with the theory of relativity—an idea still so strange that it's a byword for complexity and genius.

Even more happily, our tools of experiment, reasoning, and math give a framework for us to explore and understand the universe around us. The slow and painstaking task of discovery, making a hypothesis and testing it and remaking and testing again, has revealed something closer to the "true" nature of the universe. As with all things, we are never fully in complete understanding but always working slowly toward it. The trial-and-error nature of science sheds light on the process of knowing anything. As we explore the weird universe around us, I will also take you on a journey of discovery, knowing not just the facts as we understand them but the process of discovering those facts and the sometimes-circuitous ways to get there.

In this book, I hope to share some of the strange and beautiful parts of the universe with you. It's a weird universe, and we are lucky to live in it.

CHAPTER 1

Spacetime, the Counterintuitive Fabric of the Universe

Weird Facts: Space and time are not separate but smushed together into a weird four-dimensional construct called "spacetime." They can merge into each other—you can borrow from time to travel faster through space. The only rules of spacetime are that no matter how fast you move, the speed of light is the same; nothing can travel faster than the speed of light; and all perspectives are equally valid. From this, you can describe the nature of the fabric of the universe.

Weird Facts: Spacetime can be manipulated: masses like stars or black holes cause it to warp— that's what gravity is. And colliding masses (like two black holes spiraling inward and eventually merging) cause ripples in spacetime like a rock thrown into a pond. We can measure those ripples, but we don't yet know how to make them ourselves. Is spacetime something humans can eventually learn to manipulate? Hopefully, but we don't know yet.

To understand anything about the universe, it helps to know what it's made of. Our best understanding of the universe now (as you will see, this understanding is always under construction) tells us that we exist in a four-dimensional mathematical construct dubbed "spacetime." Spacetime is a product of Einstein's Theory of Relativity. Relativity is actually a series of equations that describe how spacetime behaves. You can provide input masses, or circumstances, and the calculations describe the curvature of spacetime. These equations were (and still are) the most precise description of gravity that we have. His first set of equations, referred to as "special relativity," deal specifically with the laws of space and time in the absence of any masses or gravity. The full "general relativity" calculations include everything and produce some pretty weird results (black holes being just one of the weirder properties of spacetime) and have a reputation as being indecipherable and something that only a genius can understand.

In our normal interactions with existence, we experience three dimensions of space. Those dimensions give us our 3D reality, where objects exist, and you can assign a location that specifies an exact point in space. The combination of latitude, longitude, and altitude gives you a precise point on or above the surface of the Earth. These dimensions of space have no preference for direction. You can move forward or backward, up and down, left and right; like a more advanced video game or a bishop in chess, you can move in multiple dimensions at once, crossing the board diagonally. In this sense, the three dimensions are related to each other—they can mix and combine. There is also no preference in spatial dimensions for any directionality—moving left is as equally valid and available for motion as right. This 3D reality is something we understand intuitively—our brains evolved in this environment, and we are good at understanding how objects move through it.

A fourth dimension exists alongside these three, but in our Earth-bound experience is always separate from the three dimensions of space. This fourth dimension is time: this inexplicable flow that we all experience. Before the development of relativity, the dimension of time was considered unchanging and uniform for all people under all conditions. Time was separated from the three dimensions of space, flowing alongside them but never mixing the way dimensions of space can combine. And unlike the spatial dimensions, time has a direction. Sometimes this is referred to as the arrow of time. You can go forward in time but never backward. The aftereffects of this one simple rule are interesting— you get the concept of causality—that things happen in an order in time and events in the past have an impact on events in the future.

Within this basic structure there are a lot of questions: Why is time different from the three dimensions of space? What causes the arrow to point only forward for time? Why are there only three dimensions of space? Why is there one dimension of time?

Some of these questions we have no answers for, others we have mildly unsatisfying answers. I can tell you that we think time points in a particular direction because of the second law of thermodynamics—the total disorder in a system can only increase over time. The term for this disorder is entropy. The second law of thermodynamics is one of the few physical laws with a time preference. Most physical quantities have no time preference and are equally valid running forward or backward. Why entropy must always increase in the universe is something we don't understand, but we can recognize immediately. A video played backward of nearly anything immediately strikes us as wrong—we know which way events should flow but we don't fully know why. Even the basic concept of why time exists is something we don't understand.

Setting our lack of deep understanding aside, we can examine that our perception of reality is of three dimensions of space that are connected to each other and one seemingly unconnected dimension of time that accompanies them. Yet, in our understanding of everyday reality, we are wrong.

The main takeaway of relativity is that time and space are not separate but are linked and related to each other. Moving through space changes how one moves through time and vice versa. This new construct of time *plus* space is defined as spacetime. A four-dimensional mathematical construct that describes the universe. Spacetime came out of Einstein's attempts to understand and explain seemingly intractable aspects of how the universe appeared to behave. The most important aspect is the nature and behavior of light. The speed of light is a constant, no matter what else is happening, and this one fact drives everything.

What's amusing about relativity is that, for all its vaunted complexity, you can understand it similarly to how Einstein came up with it. He considered thought experiments, coming up with situations like people in trains moving quickly past other people standing stationary on a train platform, people in elevators moving up and down, people falling off roofs, and tiny bugs crawling around. Ordinary situations and things that nearly everyone reading this will have experienced. Some modern textbooks use slightly different examples—planes flying and spaceships cruising through the blackness of space. But the fundamental questions Einstein asked are ones that we too can ask ourselves and get a sense of how spacetime works. One first thing to note is that all observation points are equally valid. As we discuss people on trains and people on train platforms, we note the laws of physics are the same everywhere we go.

Let's consider a person (maybe it's you) on a train, moving quickly down a track. We will describe two scenarios of things you could be doing on the train. In the first, you have a ball that you decide to throw forward in the direction that the train is moving. You perceive the ball to be traveling at a nice speed, maybe ten miles per hour. Uncomplicated. Now imagine that I am standing on a platform, and your train speeds by me the moment you throw the ball. I watch the ball fly forward at ten miles per hour *plus* the speed which the train is moving. The two speeds add up for me.

A different scenario: Instead of a ball, you have a flashlight and decide to turn it on, shining the light forward in the direction the train is moving. To you, the light leaves the flashlight moving at the speed of light. Simple and uncomplicated. Now imagine that I am on the platform and your train speeds by me at the moment you turn on the flashlight. I watch the light inside the train, and to me it is also *only* moving at the speed of light. Unlike the ball, the speed I perceive the light to travel is not the speed of light *plus* the speed of the train. It's *just* the speed of light. If, instead, you shined the light backward, in the opposite direction to the train's motion, I would still see the light traveling at the speed of light, not the speed of light *minus* the speed of the train. Regardless of how the train is moving, the speed of light is always invariant and the same.

You may see a problem here. If you, on the train, perceive the light traveling at the speed of light, crossing the length of the train car in a finite amount of time, how can I, on the platform, perceive it as also traveling at the speed of light, when I know the train is also moving and thus the light should be moving faster? The solution to this, which Einstein proposed, is that while the speed of light is always the same, everything else is relative and changeable. In the case of the train, the length of the train car appears shorter (the train is squished in the direction of motion) and time on the train passes more slowly as seen by the person on the platform. These two effects—length contraction and time dilation—occur in such a way as to result in the speed of light remaining the same. You, inside the train, are blissfully unaware of anything, assuming that your measurements and perception of time and length are correct. This is the relative part of relativity—any measurement of space and time is dependent on the conditions—are you traveling fast in a train or stationery on a platform? But the harder thing to grasp is that all measurements are equally correct. In developing his seminal paper, Einstein also described lightning strikes hitting the front and back of the train car. To you, in the train, these appear to be simultaneous. To me, on the platform, one will happen before the other. In relativity, the sequence of events also depends on your perspective.

Einstein first published his special relativity in 1905 and with it, scrambled space and time: moving quickly through space will slow time and shrink space. Einstein didn't come up with the concept of spacetime; his former professor Minkowski synthesized the formulae of special relativity into a four-dimensional spacetime. The arrow of time is maintained, but the separation between space and time is erased. The next advancement was to incorporate gravity into spacetime—special relativity deals with situations where gravity is not present (or not changing, since trains need gravity to stay on the track). General relativity moves from the special case of no gravity to one with gravity. It turns out that adding mass or energy will cause spacetime to warp and distort. What we feel as gravity is a curvature of spacetime caused by the mass of the Earth. What keeps planets in orbit around the Sun is the curvature of a spacetime well, and planets race around the edge of the well, like pennies in a funnel, curving inward along the warped spacetime. These paths are called geodesics and are the equivalent of a "straight line path" through a curved spacetime. Orbits fall along geodesics.

Einstein wrote equations that can predict the curvature of spacetime given an arrangement of mass and energy. From these equations, he correctly calculated the orbit of the planet Mercury, which had eluded and mystified physicists and astronomers for hundreds of years. Mercury is close enough to the Sun that the effect of relativity on its orbit causes small deviations that can't be explained any other way. Many others after Einstein took the equations and discovered in their depths such strange things as black holes, accelerating expansions, the curvature of light around mass, the dragging of spacetime by rotating masses, and even ripples and waves in spacetime as objects move through it.

The fundamental nature of spacetime is still a mystery to us, but we can predict its behavior and how it interacts with mass. In extreme situations—at the smallest of scales and the highest densities or both—our understanding fails, but we don't yet have a good substitute for those situations. Modern physics is continuously working to connect the physics of the universe—relativity—with the physics of atoms and molecules—quantum mechanics. Thus far, we haven't figured it out yet but given the endless amount of human ingenuity and more time, I think we will.

What this means for us: As we will see in the coming chapters, our ability to figure things out takes time and effort, but we've learned a great deal about how the universe works, what it contains, and what might happen to it. Spacetime is the framework that everything hangs on. It's a background that, for the most part, you don't notice, but occasionally, its peculiar nature takes center stage. Its weirdness and enigmatic nature are hopeful to me. The distances between stars are so vast that I don't think there is a way to travel between them in anything close to a human lifetime, no matter how good of a spacecraft or engine you build. But if spacetime is hiding secrets of other dimensions (as theorized in string theory) or if we develop a way to manipulate it directly, creating wormholes or other means of fast travel, then those distances suddenly become unimportant. The fact that we can't fully explain all of the hows and whys of spacetime gives room for the imagination, for future possibilities thus far predicted only in science fiction.

CHAPTER 2

Black Holes: A Mathematical Novelty That Turned Out to Be Real. Also, Not a Vacuum Cleaner

Weird Facts: If you're far enough away from a black hole, it doesn't matter that it's a black hole. If the Moon were replaced by a black hole, it wouldn't change anything about life on Earth except disrupting some animal behaviors that rely on moonlight (like making moonshine). From a gravitational perspective, it would be the same: tides would happen as usual. We would get weird eclipses, and there would be a strange dark shape in the sky. But otherwise, business as usual. Black holes are strange but only when you get close to them.

Having some familiarity now with spacetime, the weird fabric that makes up the universe, we turn to its weirdest implication (aside from spacetime itself): the black hole.

In popular culture, black holes are insatiable monsters, lurking hidden in space, waiting to eat everything and everything wandering by. Beware. But in reality, black holes are, while technically insatiable, incredibly well-behaved and, at least from an external perspective, predictable. Of course, this can all be true *and*, at the same time, black holes are incredibly weird. Our understanding of the nature of spacetime within them breaks down at the center of the black hole, and any indestructible observers falling into the black hole experience the universe in such a strange way that space and time appear swapped—one moves through time while space flows by. Black holes are the universe at its most bizarre and spectacular.

The concept of a black hole, defined as something with strong enough gravity that nothing can escape from it, not even light, has been around for a surprisingly long amount of time. In the late 1700s, English astronomer John Michell theorized a star so big that light couldn't escape from the pull of gravity. In this conception, the density of such a "dark star" is the same as that of the Sun, but the star would have to be 500 times larger in diameter. The escape velocity from its surface would be faster than the speed of light. Light would move away from the star and eventually slow down and get pulled back to the star. This theory was briefly popular and then fell out of favor as our understanding of the nature of light itself evolved (we don't talk much about it in this book, but light is high on the list of weird things in the universe; the behavior John Michele described is impossible with our current understanding of light). The "dark star" concept slept for 130 years, waiting for a better understanding of gravity to reemerge.

Einstein's theory of general relativity, published in 1915, gives us a framework for understanding the structure of the universe. We talked about this in the previous chapter—the fabric of the universe is spacetime, which can be curved, distorted, and rippled like a pond. The theory of relativity provides "field equations," a way to calculate the curvature of spacetime based on the locations of mass in that spacetime. It was originally thought that there would be no clean solutions to the field equations, just laborious calculations to come up with complex tables of spacetime curvature. A few months after the publication of general relativity, German physicist Karl Schwarzschild calculated a formula to predict the spacetime curvature around a single, spherical mass with no rotation, no charge, no extra bits of any kind. This "solution" to the field equations provides a useful approximation for large, slowly rotating objects like the Sun and the Earth (which each lightly distort the spacetime around them). But taken to an extreme, this solution predicts that for a massive enough and yet simultaneously small enough object (essentially a dense, heavy object), a "singularity" will form, where the terms of the field equations become infinite. The implications of a singularity in space were not explored at the time. Instead, the "Schwarzschild solution," the term for this calculation, was viewed as a mathematical curiosity, which was studied and evolved over many years (for example, physicists adjusted the coordinate system used, which made the equations simpler and more realistic), but it was never treated as if it could represent something that actually existed.

In 1967, astronomer Jocelyn Bell Burnell discovered neutron stars (to be discussed in more depth later, the dead cores of long collapsed stars held up by quantum mechanical forces). With this discovery, the field of astronomy collectively realized a collection of "gravitationally collapsed" objects exists. This realization included a reexamination of the concept of singularities and reignited the possibility that the singularities predicted by general relativity were real. The origin of the term *black hole* is not clear; it was used in 1964 but didn't become popular until astronomer John Wheeler used it after one of his talks on "gravitationally completely collapsed objects." A listener said it might be faster and easier to refer to them as "black holes," and the name has been used ever since.

HDE 226868 companion ⊂ period 5.6 days

The first astrophysical black hole ever discovered is called "Cygnus X-1" and was observed in 1965 by finding the bright X-rays that it emits. It took nearly a decade to identify Cygnus X-1 as a black hole. It is part of a binary system, with a blue supergiant star as its companion. By using the motion of the companion star, the mass of the black hole can be measured—it's about twenty-one times the mass of the Sun. This mass, which must be within a small area, as delineated by the orbit of the companion star, can only be a black hole. The X-rays are emitted by the disk of material swirling around the black hole. Because of their "black" nature, the only real way to detect a black hole is the same way Cygnus X-1 was found: to observe the motion of objects around the target black hole and determine that the mass that causes those orbits must contained in a tiny volume.

Since Cygnus X-1, we've discovered many black holes, both in the nearby and distant universe. The closest black hole we know of has a mass ten times that of the Sun and is 1,600 light years away. It was only detected based on the motion of a companion star. It's unlikely that we will ever have a close encounter with a black hole.

Most weirdly, it seems that every galaxy has a massive black hole at its center. We don't have a good explanation for why this is, and we've found some galaxies in the early universe with massive black holes that are unusually large. We think these black holes grow as galaxies merge—two galaxies with 1,000 solar mass[1] black holes merge, creating a single galaxy with a single 2,000 solar mass black hole at its center. But it would take a lot of mergers and a lot of galaxies early in the universe to build a big enough black hole.

Black holes appear to range in size from a few times the mass of the Sun to tens of billions of times the mass of the Sun. There is no limit on how massive a black hole can be. Black holes grow by eating other black holes or stars or gas from stars they've ripped apart, but this growth takes time and has certain limits. We think that there are practical limits on how large a black hole can grow, maybe to just sixty billion solar masses, maybe as high as 300 billion. For comparison, the massive black hole at the center of the Milky Way galaxy is only 4.3 million solar masses. It has a long way to go to be heavyweight champion.

---

1   A solar mass is a convenient unit to work with, given how large things in space can be. It's equivalent to the mass of the Sun. Helpful, right? In more ordinary terms, it has a mass of 1988400000000000000000000000000 kg or 1.9884×10^{30} kg. Now I'm sure you see why we shorten it to one solar mass. Sometimes we will also use Jupiter masses (1.899×10^{27} kg), which is about one thousandth the mass of the Sun. An Earth mass (5.97×10^{24} kg) is even smaller—300 Earths can fit into Jupiter. 1,000 Jupiters can fit into the Sun. Throughout this book, we will sometimes use these masses rather than listing things in kg. In the example above, a black hole of 1,000 solar masses has as much mass as 1,000 of our Suns. Some stars will have more mass than our star, some will have less, but the unit is based on our little Sol.

How to Make a Black Hole

A typical black hole forms in the death of a star. When massive stars run out of fuel for fusion, their cores collapse under the intense force of gravity. For stars larger than a few times the mass of the Sun, the core will be condensed into a neutron star, held up by neutron degeneracy pressure—effectively the core becomes one giant neutron, where the repulsive force of protons pushing against each other provides the check against gravity. For more massive stars (probably about twenty to thirty times the mass of the Sun or larger), neutron degeneracy pressure can't support the mass of the core, and the core collapses into a black hole. In both cases, the remainder of the star explodes outward, having bounced off the collapsing core, creating a spectacular supernova explosion. These explosions are an engine of creation, creating heavy elements and spreading material into the galaxy to form new stars and planets. The black hole left over from this explosion remains behind, enigmatic and hidden in the galaxy. We think there are 100 million stellar-mass black holes in our galaxy, but most of them are impossible to observe. In the case of Cygnus X-1, we could only detect it because of the companion star. In many cases, without a companion or some other interaction drawing attention to it, the army of stellar-mass black holes in the galaxy will remain forever unobserved. We know statistically that they are there but don't have their exact locations and qualities. This makes studies of black holes challenging since they are nearly impossible to observe on a large scale.

There may be other ways to make black holes aside from collapsing stars. Most of these methods rely on the weirdness of the early universe and, at this point, are all theoretical. The first way creates "primordial black holes," which would have formed in the first moments after the Big Bang (less than a second after the start of time). These black holes would form from random fluctuations in the density of matter during a period of massive expansion known as inflation and would almost accidentally create a black hole. There is a huge debate about how massive these primordial black holes could be (from teeny tiny to thousands of times the mass of the Sun), how long they could survive (small black holes evaporate away, but large black holes last for a long time), and if they are the seeds for galaxies to form (maybe, if the black holes are large). These primordial black holes are also theorized as a possible explanation for dark matter (a weird form of matter that we don't understand), since, as we just learned, black holes that don't have some type of companion star are difficult to detect. More observations of the early universe (like with NASA's James Webb Space Telescope) may help to either prove or rule out the existence of primordial black holes.

Another way to make a black hole in the early universe is via direct collapse. Again, this is theorized and not yet proven. As we will see in later chapters, stars collapse out of dense, cool clouds of hydrogen. We know that in the current time, this process involves fragmentation—where the entire dense, cool cloud of hydrogen splits into smaller chunks, each chunk making a small-ish star. In a direct collapse formation, the cloud doesn't split (likely due to the lack of heavy elements and other sources of instability in present-day clouds that didn't exist in the early universe) but keeps collapsing down into one massive object.

The cloud never becomes a star (although a similar process likely created the first generation of stars, all of which were stunningly massive) and collapses into a black hole. The hydrogen skips the star phase, fusion never even has time to push back on the force of gravity, and the entire mass of the cloud exits immediately into a black hole. This process could make black holes around 10,000 times the mass of the Sun, and again, these could be the seeds of the supermassive black holes at the centers of all galaxies.

There may yet be other ways to create black holes. We only barely understand spacetime and how it can be manipulated. One lingering question about black holes is how every galaxy seems to have a supermassive one at its center, especially in the early universe. We don't have a deep understanding of how such large black holes can be built.

What Happens Inside a Black Hole?

This is the real cosmic question about black holes. Interestingly, the field equations for a black hole don't blow up until the center of the black hole. The black hole's boundary is defined as its event horizon (although non-black hole objects like the universe can have event horizons). In the case of a black hole, this surface is colloquially known as the point of no return—where the curvature of spacetime is so great that even light, traveling at the speed of light, cannot escape. The event horizon is not a surface—it's more of a mathematical boundary that shields the singularity of the black hole from the outside universe. Things don't crash into the event horizon or get absorbed by it; there is no marker. In fact, anything traveling across the event horizon won't realize they have passed this irreversible Rubicon.

From a relativistic perspective, an outside observer can never know what happens within the event horizon of a black hole. An observer outside of the horizon can never see what is happening in the singularity, and these shielded singularities seem necessary in general relativity. This is partly to preserve fundamental concepts like causality—the relationship between cause and effect. Someone watching a (doomed) friend in a spaceship enter a black hole (ideally watching from a safe distance) will never see their friend cross the event horizon. They could watch until the heat death of the universe (more on that later) and only see their poor friend approaching the event horizon but never crossing it. Their friend flew by it quickly, but the outside observer can never see this happen.

The true definition of the event horizon is trickier and reveals more about what happens within the black hole. For objects orbiting the black hole, there is a last stable orbit outside of the event horizon. This "last stable circular orbit" is the closest to the black hole that anything can orbit and not get pulled in. A photon can keep this orbit at 1.5 times the Schwarzschild radius for a nonrotating black hole. This region is nicely called the photon sphere. Objects passing within the last stable circular orbit don't necessarily enter the event horizon, but they need to have enough energy to fly by the black hole and keep going past it. Anything in an unstable orbit or without enough energy will end up on trajectories (called geodesics) that end by passing through the event horizon and eventually land at the singularity.

Interestingly, no possible configuration would result in a perfect orbit at and around the event horizon. The event horizon itself is a "null," no orbits or paths can be on the event horizon; they just pass through it. As you make it through the event horizon, all motion is in the direction of the singularity, deeper into the hole. This is true for light and for any other object entering the black hole. Within the event horizon, spacetime becomes scrambled— no matter which direction you move in, you are always moving toward the singularity. All motion becomes reduced to a single direction. We experience something like this in our perception of time—the arrow of time moves only forward, no matter how you move in space. In a black hole, any motion forward in time equals a motion toward the singularity. The flow of time compels all objects toward the center of the black hole. One analogy for this is once you go over a waterfall, no matter what you do, you are forced to travel with the flow of water and reaching the bottom is inevitable. Entering the black holes makes the singularity an inevitability. The singularity itself is possibly removed from spacetime—not having a "where" or a "when." This is the part where the field equations are hopelessly broken, and we are at a loss.

I still haven't given you a good answer for what happens inside a black hole. That's partly because we don't know the answer. As far as we can guess, once an object or photon reaches the singularity, it gets destroyed and contributes to the mass and energy of the black hole, which grows a little (making the event horizon slightly larger) while the singularity gets more massive but does not grow in size. The density of the singularity, smushed into what is effectively zero volume, is infinity. Thus, the problem that the black hole reveals: dealing with infinity of anything is difficult, and infinite densities in a volume with technically zero size are doubly difficult. Small volumes are generally the province of quantum mechanics, not relativity, thus physics has searched for a theory of quantum gravity to couple these two opposed concepts. If such a model could be found, it would combine the behavior of large masses (governed now by general relativity) and the small volumes of the atom (ruled by quantum mechanics). Other theoretical physics work is driven by this same conundrum. The weirdness of black holes has been an incredible font of inspiration for physicists trying to figure out how the universe works.

What this means for us: If you are worried about black holes engulfing the Earth and ruining your day, I am here to assure you that you don't need to worry. We will never be vacuumed by a black hole. One thing to consider is what happens if the Sun became a black hole (but also don't worry—it won't ever). The entire mass of the Sun would be condensed to within a three-kilometer radius and, of course, it wouldn't shine in the sky anymore. Setting aside the massive problems on Earth from the entire collapse of the food chain, the actual gravitational effect would be precisely nothing. The Sun is still the same mass, just denser. At the distance which the Earth orbits, there would be no change in the curvature of spacetime, and the Earth would continue to blithely orbit as it does now. The only time things become dangerous for stars, planets, or spaceships is if they wander too close to the black hole, entering orbits with no escape. In some sense, it's a more extreme version of what happens to things like meteors on Earth or comets around the Sun. Some orbits pull you into the object and that's that. The same is true for black holes, just taken to the absolute limit.

While black holes are interesting, and the nature of the singularity inside each one is currently mysterious (Nobel Prize for whoever figures it out), from the outside, they are well-behaved, completely predictable, if not a little sinister, cosmic neighbors.

CHAPTER 3

Space Is Actually Quite Large (Larger Than You Think)

Weird Facts: Space is so large that the two satellites leaving the solar system (Voyager 1 and 2), traveling at nearly 35,000 miles per hour, will need 40,000 years to reach the next star. Even light, traveling at the speed limit of the universe, takes 100,000 years to cross the galaxy and millions or billions of years to reach other galaxies.

We've talked about the most compact and dense and condensed space can be (the singularity of a black hole), so for some perspective, we should consider the opposite—just how large space can be.

While the title of this chapter might seem self-evident, just how big space is is a difficult quantity for humans to wrap their minds around. Limited as we are by the realities of life on the surface of the Earth, even imagining the distance from the Earth to the Moon is challenging (and we've traveled there). Conceiving of any other distances is difficult and maybe impossible for our brains, evolved as they are for the Earth's surface.

One way to think about the vastness of space is to consider how long it takes light to travel across it. Starting small, consider Earth. Light travels at 299,792,458 meters per second or 186,000 miles per second. From New York City to Los Angeles is about 2,446 miles, as the crow flies. Flying on a normal commercial airplane, you can cover that distance in about five hours, with reasonably good routing from air traffic control and a nice tailwind. Light can cover that distance in 0.013 seconds. A little over one-hundredth of a second, this is roughly a million times faster than you can travel. Light from the Sun arrives at Earth 8.3 minutes after leaving the Sun's surface, covering nearly ninety-two million miles in those 8.3 minutes. Light from the Sun takes 13.4 minutes to get to Mars and forty-three minutes to get to Jupiter. Neptune, the farthest planet from the Sun, is 2.8 billion miles away from the Sun, and it takes light four hours to travel from the Sun to Neptune. These distances are so great that even light, the fastest thing there is, can take hours to cross the solar system.

I pause here to make a note about the use of "light years" as a distance scale. It may be confusing at first—shouldn't something that has the word years be a unit of time? A single light year is a unit of distance that we've created that's calibrated by the length of a year on Earth and how quickly light moves. So, a light year is the distance that a photon will cover across space in a year. It is specifically a length (one light year is about six trillion miles, or 6,000 billion miles), but it also gives you a sense of the travel time for a photon. If I tell you something is two light years away, it's twelve trillion miles away but it would also take light two years to bridge the distance. But for you, a space traveler not traveling at the speed of light, it would take many more years to cover a two-light-year distance. Thinking about the universe in "light years" is essential because the units of length we developed here on Earth (miles and kilometers) are so tiny that they are useless in space. On the scale of the universe, a light year is still tiny (we measure the universe in billions of light years), but it's a pretty good measure for the solar neighborhood and galaxy, so we use it everywhere.

Our solar system is at the smallest scales of distance that we can experience. Pushing our light bubble outward, let's explore farther beyond our tiny neighborhood into the rest of the galaxy. The distances between stars in the galaxy are so vast that it takes four years for light from our Sun to reach the next closest star, Proxima Centauri. Within a radius of about twenty light years from the Sun, there are only 131 stars, and only twenty-two are bright enough to be visible without a telescope. That's 255 cubic light years per star, a massive volume for each of these tiny objects. Space is empty and large.

Beyond our little solar neighborhood, the center of our Milky Way galaxy is 26,000 light years away from the Sun. Light that was emitted from the Sun when humans were first crossing into North America 25,000 years ago is only just now reaching the center of the galaxy. We are on a spiral arm of the Milky Way, about halfway out. The full extent of the Milky Way is 100,000 light years across. Light traveling to the next closest large galaxy, Andromeda, covers 2.5 million light years. Light traveling to the *most* distant galaxy we've observed, JADES-GS-z14-0, traveled for 13.4 billion light years to reach us (although a weird quirk of the expansion of the universe is that JADES-GS-z14-0 is currently 32 billion light years away from Earth and is outside of the visible horizon—i.e., the light it's emitting right now will never reach us. We will never be able to observe it as it currently appears).

So we've gone from light minutes needed to travel between the Sun and the Earth to billions of light years needed to travel to the edge of the known universe. If the expansion of the universe continues, eventually distances will be described in trillions of light years and more. These lengths, and the empty space found between stars and galaxies, are so massive that it's hard to convey them. Even the discussions of time that I've tried to make are tricky—what does a billion years mean to a human who will live only 100 of those years?

One other way to think about distances in space is to leave photons behind and consider how long it might take humans in a spacecraft to travel at cosmically puny speeds but still fast for a human being. We can again start close to home—going from the Earth to the only other solar system object that humans have landed on: the Moon.

Traveling on an Apollo spacecraft (rides are tragically no longer available), it takes a person roughly 3.5 days to get to the Moon. That seems good—we always think of the Moon as being close to us. But in every graphic of the Earth and Moon together that you've likely seen, the scale is wrong. The Moon is shown to be closer to the Earth, making the apparent distances seem smaller, a more bridgeable gap. But we can show a true image of the Earth and the Moon together, taken by interplanetary satellites like the one here taken by OSIRIS-Rex.

NASA/GSFC/University of Arizona

In this view, both the Earth and the Moon are small, fragile spheres against a vast blackness, a callback to the first picture taken of the Earth as a complete sphere against the blackness of space. An echo of the picture of Earth from the outer solar system taken by Voyager: what Carl Sagan famously called "the pale blue dot." Both planet and Moon appear puny compared to how far away they are from each other, and the Moon appears even tinier than we usually think of it. This is a huge distance for a Moon that is only a mere 250,000 miles away. In a cosmic sense, it is directly on top of us. Close enough to create the strong gravitational effect of the tides and appear as the second brightest object in the sky (after the Sun).

The Moon is close enough that it's relatively easy to reach—with 1960s-era technology, a liquid-fueled rocket, and gravity, you can get there in just a few days, hang out, and come back. A week-long trip to our lunar friend. That's faster than a 1600s sailing ship to cross the Atlantic.

The next closest solid ground to visit (Venus and Mars) are both an order of magnitude longer distances. Unlike the Moon, the distance between Earth and every other planet in the solar system can vary, depending on where everyone is in their orbit. Mars, the most typical destination for things from Earth, can either be as close as 56 million kilometers (roughly half the distance between the Earth and the Sun), or as far away as 400 million kilometers (eight times as far. This maximum distance happens when Mars is on the other side of the Sun from Earth). The closest approaches happen about every two Earth years.

Getting to Mars takes a lot of energy—the orbit of Mars is farther from the Sun and the orbital speeds are different, and your spacecraft will naturally slow down as it travels away from the Sun (the energy in the speed of your spacecraft gets used up to counteract the effect of gravity, the way a ball thrown in the air will slow down as it ascends). Because of these factors, a significant amount of energy is used to just move between the orbits. The lowest energy orbital maneuver is called a "Hohmann transfer orbit," which takes nine months to get to Mars. Couple that with nine months back, and the fact that the Hohmann transfer only works when the planets are aligned every two years, and the total time for a round trip to Mars is thirty-four months. You either have to stay for sixteen months to wait for the next alignment or figure out a different, more energy-intensive way back. Other orbital maneuvers can get you there faster but usually require more energy. And because of the two-year alignment, you'll need to leave Mars almost immediately or stay there for a two-year cycle. Venus is closer to Earth at its closest approach, just forty million kilometers, and is slightly easier to get to from an orbital dynamic standpoint, taking just under four months. But then you've at Venus, whose surface is a hellscape of intense heat, pressure, and a sulfuric acid atmosphere, which is why we spend most of our time sending things to Mars instead of Venus.

Traveling to any other large solar system object, we need to carefully consider the destination and the timing of the trip. If you want to get somewhere far, but don't care how long it takes, there are many options. If you are sending a human somewhere, you (probably) care how long it takes to get somewhere—ideally your human would still be alive when reaching their destination (but maybe you are less concerned for their well-being and just want to get them as far away as possible; no judgment from me). In the case that you want your human to still be alive at the end of the trip, your options are more limited. Traveling to Jupiter, for example, will take you at least a year and a half but sometimes longer. That is a perfectly human timescale, although Jupiter itself is a harsh destination. Trips to Saturn, the second largest planet, are even longer—a few years to as long as ten years, depending on what Jupiter is up to. A smart traveler will use what's known as "gravity assist" whenever they can.

A gravity assist is a tricky orbital maneuver, invented at JPL in the '60s, which uses the gravity of the planets to speed up a spacecraft and fling it in a desired direction. A spacecraft will do a "fly by" of a planet or moon, approaching it in such a way that it can steal some of the planet's momentum and energy. After the close encounter with the planet, the spacecraft will depart, moving faster and in a slightly different direction than when it arrived. This is a little like baseball—a pitch thrown gets hit by the bat and ends up moving in a different direction and adding twice the bat speed to its speed. Effectively you bounce your spacecraft off a massive planet and it's suddenly moving much faster. This technique has been used many times on probes like the Voyager missions, and the larger the planet or moon you use, the bigger the speed boost you get. That's why Jupiter is by far the preferred gravity assist partner, being by far the largest planet. Of course, this requires a good alignment between Jupiter and your destination, which isn't always convenient.

If you don't care about how long it takes to get somewhere and you'd like to use as little energy as possible, a set of calculations have generated trajectories for moving around the solar system known as the "Interplanetary Transfer Network." While this sounds like something from *Star Trek*, it's a real way of moving around the solar system. The network is like the gravity-assist technique in that it relies on the motions of planets and moons, but in general is slower and more complex. Normal orbital mechanics usually only considers the effect of one large mass at a time—if you are using Jupiter to do a gravity assist, you calculate the effect of Jupiter on your spacecraft; the Sun and other solar system bodies don't matter. But for the ITN, you care about multiple bodies—the Sun *and* the Earth *and* the spacecraft, for example, or the Earth *and* the Moon *and* the spacecraft. In this case, there are three bodies to deal with. In orbital terms this makes calculations more complicated but also creates nodes of gravitational stability between the bodies. The ITN uses these nodes to allow for nearly energy-free travel between objects. Experts who have calculated the possible trajectories describe a system of metaphorical "low energy passageways" between solar system objects; sometimes it's described as a series of tubes (like the internet). Of course, the tubes are not real but describe twisting and winding paths through space that allow for transfer between solar system objects without propulsion. We think this is how slow-moving asteroids can travel to the Earth—they happen to be pushed into a particular trajectory that will eventually lead them to Earth. Using the ITN, it could take decades to get to Mars (which would be a months-long trip with a faster trajectory), but it wouldn't cost you any fuel. This type of network could also exist between the Sun and other stars as well but again would be enormously slow.

Beyond the solar system, the distances are so vast that even calculating the time to travel there feels like a futile exercise. Voyager I, launched in 1977, is the most distant human-made object. Currently it's twenty-four billion kilometers away from the Earth, or 163 times the distance between the Earth and the Sun. It's been traveling for over forty-five years and will take roughly 40,000 years to reach another star (the star is called AC+79 3888, only seventeen light years away from us). It happens to be pointed at this random star, but even if it were traveling to the closest star to us (Proxima Centauri, four light years away), it would still take 10,000 years to get there. Definitely not a human-friendly trip.

Thinking about travel to even more distant places, like the center of our galaxy or the next closest galaxy, it's convenient to switch our thinking back to the speed of light, as we explored at the beginning of this chapter. This is because while it takes Voyager

40,000 years to get to AC+79 3888, it takes light a short and sweet seventeen years to get there from Earth. Light provides a useful measuring stick for all things galactic and intergalactic, and even then it will take light tens of thousands of years to cross the galaxy. Our Milky Way galaxy is about 100,000 light years across. In the time it takes Voyager to get to AC+79 3888 (40,000 years), a photon emitted at one end of the galaxy will be almost, but not quite, halfway across.

What this means for us: When one considers interstellar or intergalactic travel for humans, it's clear that even moving at the speed of light isn't fast enough (that's even leaving behind the current impossibility of getting a spaceship moving close to the speed of light). Science fiction has developed techniques to circumvent this—the warp drive, wormholes, and other dimensions where fast travel is possible. It remains to be seen if science will ever discover a way around these impossible, massive distances. Energetically speaking, accelerating a spacecraft to the speed of light is impossible. But before the discovery of the gravity assist, scientists thought that travel to the outer solar system was energetically impossible—it would take too much energy to boost even a tiny spacecraft to such a distant orbit. But a deeper understanding of gravity, orbital dynamics, and propulsion opened the solar system to exploration. Maybe one day, our understanding of spacetime and the fabric of the universe will enable future explorers to make a light-year distant trip in just a few minutes. Until then, we must contend with how tiny our world is, floating in a vast, empty, and immense universe.

CHAPTER 4

Stars, the Universe's Response to Gravity

Weird Facts: On a dark night, away from city lights, you can see about 2,000 stars. Each one is an ongoing fight between the forces of gravity and the outward pressure of light. Each star is a temporary waypoint before gravity eventually wins and the star dies. Technically the fusion that powers every star shouldn't happen, yet it does.

The defining feature of the night sky is the stars. To an early human, the bright object in the sky during the day may not have held any relation to the tiny pinpricks of light that illuminate the night sky. From our vantage point on Earth, we see a sphere of stars that rotates around us over the course of a night and over the course of a year. Early humans used the appearance of stars at sunset and sunrise as a rudimentary calendar system. The Greeks referred to the distant, seemingly unmoving stars as the "celestial sphere." Stars are the main parts of all galaxies that we see. Stars are the home of every planet we know of. The star is the fundamental unit of the universe.

But what is a star? At its most basic definition, it's the natural result of the ever-reaching pull of gravity. Any universe that contains gravity should form a condensed object. Gravity pulls things together and acts over long distances. This means even a tiny over-density (a region of space with a few more atoms than average) will grow, becoming denser and more massive. Eventually these over-densities condense, heat, and ignite. Something too over-dense will become a stranger object—a white dwarf, neutron star, or a black hole. But those lie in wait for us in a later chapter.

Λ Star Is Born

In our universe, the most abundant element is hydrogen. Imagine a volume of space containing an even distribution of hydrogen atoms—a perfect diffuse and smooth dispersal of atoms. In this even grid of hydrogen atoms (picture an almost crystal-like structure), there is no preferred direction for gravity to pull things toward. All atoms exert a tiny pull on all other atoms, and, on balance, everything cancels out, leaving a static and never-changing arrangement (for this thought experiment, the edges of this volume of space, of course, present a problem, but for now, let's consider it to be infinite). But this arrangement is hugely unstable: gravity is always working. To see just how unstable, let's add one hydrogen atom to the grid so it's close to another one. In this evenly spaced sea of hydrogen atoms, one region has a little more mass compared to the rest. That tiny over-density will now exert a slightly greater pull on all the other atoms around it, slowly attracting them toward it. What started off as two atoms in proximity will quickly snowball into many. The inexorable pull of gravity, once started, cannot be undone. Over time (and the length of time this condensation takes can be variable, from a few tens or hundreds of thousands of years to ten million years or more), the atoms collect into such a large aggregation that it's more practical to talk about them in terms of larger objects. We can compare the mass of our condensing cloud of hydrogen to the mass of the Sun—stars eventually have a range in sizes from one-tenth of the mass of the Sun to hundreds of times the mass of the Sun. As the cloud gets denser, the atoms farther away in our thought experiment are moving closer to the center of the cloud, toward that original over-dense point. The cloud is said to collapse like an explosion running backward in time.

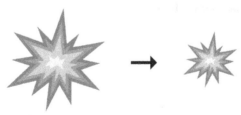

The center of the cloud grows in mass, temperature, and density, and eventually this situation causes an ignition.

The collapsing cloud of hydrogen is in an interesting race against time. The more material that can be accumulated at the center of the cloud, the larger the star will be. But the material at the center is also heating—the energy of collapse needs to go somewhere and will cause the hydrogen atoms that were so cold at the start of this process to become hotter. The heat eventually creates a protostar—not yet a star, but no longer a collapsing cloud. The outward pressure from the motion of the hot hydrogen atoms prevents the protostar from collapsing further, counteracting the pull of gravity. This is the first instance where gravity has been temporarily halted. The star is essentially forming from the inside out; the protostar will continue to grow in the outer layers, but the inner region—a protocore, if you will—is in a quasi-equilibrium. The protostar still grows, but the bulk of its mass is already present. More slowly, the protostar condenses and heats, causing phase changes to the atoms that compose it. Molecules get broken, hydrogen becomes ionized, and the protostar can cool more efficiently. This cooling, somewhat counterintuitively, is a critical part of the evolution of the protostar. By cooling, the thermal pressure opposing gravity is no longer as strong, and gravity reasserts itself. The protostar shrinks, becoming denser, smaller, and hotter. This process happens enough times that, finally, it shrinks enough that the core of the protostar achieves the ignition of hydrogen. A star is born.

Your Example Is Not Real

Now wait a minute. My star formed out of a perfectly smooth, perfectly even mass of hydrogen atoms that continues infinitely in all directions. You may ask yourself, how often does one find that perfect, smooth arrangement in the universe? Of course, the answer is basically never. Even in the early universe, the first stars formed out of hydrogen and helium, both created in the Big Bang. The distribution of atoms was not perfectly smooth but we now believe it was slightly irregular. These irregularities on both star-sized scales and galaxy-sized scales are the seeds of all structures in the universe—from stars to galaxies to clusters of galaxies all the way to superclusters and the cosmic web.

In the early universe, the first stars that formed were likely massive—huge monsters that formed quickly and died even faster. We believe if only hydrogen and helium are present (and no other heavier elements), then the stars that form out of that pristine material will be much larger than the stars produced now—potentially maxing out at 1,000 times the mass of our Sun. A star this huge will die in a few million years, ending its life in a spectacular supernova explosion. Even the largest stars formed now are, at most, a few hundred or so solar masses. These modern massive stars will still die within ten million years of ignition. In contrast, our Sun will last for ten billion years. The smallest stars (about one-tenth of the mass of the Sun) will likely survive for ten trillion years, long after the Sun and all of us will have been a distant memory.

A small digression to talk about comparisons. It's hard to conceptualize how long stars can live. The smallest stars forming today will outshine and outlast every star that you can see in the sky. The closest star to us, Proxima Centauri, is this type of small star, called an M Dwarf. It will live for about four trillion years. We can only see it because it's just four short light years from Earth. But Proxima Centauri will be around long after the Sun has died and destroyed Earth as it dies. Proxima Centauri will be around after the Andromeda galaxy has collided with our Milky Way (four billion years from now). Proxima Centauri will be around when the accelerating expansion of the universe means that no other galaxies will be visible outside of our local cluster. The "observable universe" will cease to exist to any observer, yet Proxima Centauri will only be entering middle age.

For a human lifespan, even the million-year ages of the shortest-lived stars represent an eternity. The lifetime of one massive star formed in the early universe, a short million years, is longer than all of human history. These early, massive stars were key drivers of change in the early universe. During their brief lives, they put out a huge number of photons and tremendous amounts of energy, shaping and changing the baby galaxies they formed in. In their deaths, they formed the first instances of all elements heavier than lithium—creating the primary elements that make up the Earth—carbon, oxygen, silicon, and nitrogen. These elements and everything aside from hydrogen and helium, were forged in the heart of a dying star. By creating these heavier elements and sending material out into the galaxy in their explosive deaths, these stars created the seeds of the next generation of stars and eventually the stars we see today. Our Sun has reasonably high amounts of these elements, which is partly why it's the size that it is and why there are planets around our Sun. Those first, early stars had zero planets because there was nothing to form planets from. A universe with only hydrogen and helium can't make a rocky planet like the Earth. In any event, those early stars died so quickly that the star barely had time to form before it exploded, never mind making anything else.

How Do Stars Form Today?

Around us in the galaxy right now, stars form out of more complex beginnings than that hypothetical massive first star. Giant molecular clouds, dense complexes of molecules of primarily hydrogen and traces of other elements, in our galaxy are the seeds of new stars today. These stars are all smaller than the first stars, with most less massive than the Sun. In any star-forming complex, most stars formed will be tiny—so many more of the trillion-year shrimps than the million-year giants.

But like many things, the true process of star formation in these clouds is still uncertain. We believe there is a complex interplay of slight over-densities, variable velocity and wind structures within the cloud, magnetic fields within the cloud, turbulence, and gravity, which combine to trigger star formation in the modern era.

Stars are the lifeblood of any galaxy. A young, vibrant galaxy is defined as one that is actively forming stars, one that still has dynamic arms, growth, explosions, and action. There are galaxies where no new stars form, and many of the stars left in a galaxy are small, dim, and relatively unchanging. In a universe most notable for its dynamism and constant change, these static galaxies are a study in contrast.

Nuclear Fusion Defines the Star

Hydrogen ignition defines a star. At high enough temperatures and densities, four hydrogen atoms get transmuted into a single helium atom. This process, technically a type of alchemy (converting one element into another), also emits two highly energetic gamma-ray photons. The energy to create these two gamma-ray photons comes from the nearly infinitesimal mass difference between four hydrogen atoms and one helium atom. The four hydrogen atoms together weigh the slightest bit more than the resulting one helium atom. The difference is 0.02870 united atomic mass units (u, technically one twelfth of a free carbon-12 atom at rest in the ground state). In grams, this is 4.7×10^{26} (or 0.00000000000000000000000000047) grams. This amount of matter gets converted to energy following Einstein's most famous formula, $E=mc^2$. E is the energy liberated by converting m mass into energy. c is the speed of light, and when squared provides a huge multiplier.

Why Is There Even a Mass Difference?

This quirk of reality is the reason that anything exists at all. When a nucleus is formed, it always has a little less mass than if you counted all the protons and neutrons that went into it. This mass is the "nuclear binding energy"—it's the energy liberated by binding the subatomic particles together. The bound nucleus has less energy than the free-floating particles. Our entire universe, and certainly all life on Earth, relies on this mass difference. If this wasn't the case (if creating a nucleus wasn't energetically favorable), no nuclei would form and there would be no atoms. The fundamental reason *why* this is is tricky. If this wasn't the case, you wouldn't be reading this book and so we are somewhat forced to live in a universe where a nucleus can form and be energetically stable.

Okay—so a tiny amount of energy is liberated in each of these fusion reactions (*fusion* because we are fusing hydrogen into helium, fuse). This energy keeps the star from collapsing under the force of its gravity. The gamma-ray photons liberated by each fusion reaction push outward, providing a pressure that holds the star up. The continued existence of the star is predicated on the continuation of fusion. The star *must* fuse to keep from collapsing.

When the star runs out of hydrogen to fuse, which can happen for a variety of reasons, it loses its best, most reliable source of outward pressure and can no longer support itself against gravity. It doesn't die instantly, but it has taken the first step down a path that only ends in its death.

Stars Only Shine Because of Quantum Mechanics

If we followed the laws of classical physics, stars wouldn't shine. Even within the core of our Sun, the temperature and pressure aren't high enough for nuclear fusion to occur. The positive charge of the proton repeals other protons, and the whole process of making helium is that you get four protons to join in a harmonious proton mush (a.k.a. helium). Under ordinary conditions, two positively charged particles repel each other (try some experiments with balloons and wool and feathers at home to explore electric forces), so to get them close enough to each other to join into a nucleus, they need to be moving fast and be under tremendous pressure. The temperatures inside of stars aren't high enough to overcome this repulsion—called the Coulomb barrier. But we know stars shine and fusion is the fuel. The key to this is quantum tunneling.

This is a weird bit of the universe that says if there is a barrier that *should* be insurmountable to a particle, it isn't always insurmountable. Quantum tunneling relies on the fact that the exact amount of energy a particle has isn't perfectly known. There is always a blur around both the position of any one particle and its motion/energy. This lack of precision is one of the hallmarks of quantum mechanics—the behavior of a particle is all about probabilities, not definitive values. Thus, our proton trying to get close to another proton to fuse has some tiny possibility of making it somewhere it shouldn't have enough energy to go. Given that there are a lot of particles in the Sun, for example, there is always a chance that a proton ends up right where it shouldn't be, close enough to another proton to start fusion. So, protons effectively tunnel under the Coulomb barrier and fuse when they shouldn't have enough energy to do so. It's kind of amazing that it happens.

Let's explore for a moment if quantum tunneling wasn't a thing—if somehow the universe was organized differently. The immediate effect would be that stars need to be hotter and denser before starting fusion to support themselves against gravity. Our Sun would shrink, and its core temperature would rise until a temperature and pressure are reached that allows protons to overcome the Coulomb barrier by brute force. Interestingly, the Sun probably isn't massive enough to ever reach the temperature required to fuse in a non-quantum environment. Stars would only shine if they had masses above around five solar masses. Below that, a star like the Sun would reach a different state of matter (held up by the pressure of electrons; more on that later) and never achieve the high core temperatures needed to ignite fusion. Luckily for us, we live in a weird universe with quantum tunneling to cheat the Coulomb barrier and make energy without needing to do fusion the hard way.

What If There Weren't a Mass Difference?

Sometimes it's fun to consider other universes (I'm sure our ideas of fun are the same, right?)—what if these small but important numbers (the mass of an electron versus a proton, the fine structure constant, etc.) were slightly different? Earlier in this chapter we discussed the binding energy of a nucleus. If the binding energy were a different value, stars would still form but they might be slightly brighter or dimmer or live for longer or shorter lives. The existence of the binding energy means the star will form eventually and temporarily halt its collapse against gravity during its lifetime. If the binding energy were zero, or even positive, there would be no energetically motivated reason for hydrogen to fuse into helium, and thus, no outward pressure against gravity by making light. No shining stars. A dark sky in those universes. Lucky for us, we live in a universe that seems balanced on the edge of a knife— just right for us to observe everything that's going on.

Interestingly, it took us approximately forever to figure out what powers a star. Early Greek thinkers (Anaxagoras and Aristarchus) proposed the stars were objects like our Sun, and they were so unbelievably far away that they appeared much dimmer. It wasn't until the 1800s that Friedrich Bessel reliably measured the distance to a star called 61 Cygni and proved stars were incredibly distant compared to objects in the solar system. A Jesuit astronomer, Angelo Secchi, conducted extensive observations of the Sun and thousands of stars in the late 1800s and realized they were similar in many ways.

The source of the Sun's power, hydrogen and its fusion, was first discovered by Cecilia Payne-Gaposchkin, an astronomer at the Harvard College Observatory. In her PhD thesis, she combined her understanding of neutral atomic emission lines, a new theory by Indian physicist Meghnad Saha, and the extensive spectral catalogs of the Harvard Observatory, and realized that the spectral signatures in stars were indicative of hot hydrogen. Her thesis, called "Stellar Atmospheres," was published in 1925, and argued that stars were made of hydrogen, in an abundance a million times more than other elements. She also argued that helium was prevalent in the Sun. The existing assumption was that the Sun and the Earth were made of the same composition, and Payne-Gaposchkin's assertion otherwise, despite being supported by data and clear evidence, was not well received. She had to rewrite parts of her thesis related to this to receive her PhD. Not four years later, her thesis advisor published a paper that argued exactly what she had written and been forced to change. Not the first instance of human biases setting back the progress of science and discovery.

Thanks in part to Dr. Payne-Gaposchkin, we know now that stars are mostly hydrogen, the simplest and most abundant atom. Later work, incorporating the new fields of quantum mechanics and nuclear physics, helped to prove that the energy source powering the Sun wasn't combustion or gravitational energy but nuclear fusion.

What this means for us: Our entire ecosystem relies on light and energy from the Sun. Without it, we would all die quickly. The production of that light is dependent on the weird quirks of quantum mechanics. We're lucky we live in a universe that can make stars. While it's true that, at a basic level, all stars are essentially the same, we've found a huge range in characteristics in stars in our galaxy and universe.

CHAPTER 5

The Biggest, Fastest, Smallest, Weirdest Stars

Weird Facts: Stars come in a wacky and wild assortment of colors, sizes, temperatures, and fates. Despite being driven by fundamental forces (gravity, mostly, with a sprinkling of quantum mechanics), stars are hugely varied and interesting. Stars are beautifully colored (although never green) and fabulously complex.

Stars, as we've established, are the universe's response to gravity. A force that reacts to matter and always pulls inward eventually creates a star. In some sense, stars are simple—massive collections of hydrogen, with enough pressure and heat at the center to fuse hydrogen into helium. Technically, that's it—the trait that determines nearly everything about the life of a star is its mass.

Yet the universe of stars contains a huge array of types, colors, interior structures, lifetimes, and fates. For such a seemingly simple object, they are astoundingly complex.

Stars Come in More than Just One Color

Colored stars are visible to the naked eye from nearly anywhere on Earth. If it happens to be wintertime when you read this, an entire color wheel is in the sky for your enjoyment. This color wheel is called the "Winter Hexagon"—it's not a true constellation like Orion or Gemini, but instead what we term an asterism, any grouping of stars. Asterisms are informal and not part of the official eighty-eight constellations that cover the sky.

The winter hexagon is made by following several bright stars high in the sky in mid-winter. We start, as nearly all winter stargazing does, with Orion. In addition to the distinctive three stars of the belt, Orion contains two bright stars. Rigel, Orion's knee, is a bright white/blue star. Move to the left and you'll see Sirius, in the constellation of Canis Major (the Big Dog), the brightest star in the night sky, also white/blue. Continue the circle, moving up and clockwise and you hit Procyon in the constellation of Canis Minor (the Little Dog), whiter and less blue; and, onward around the wheel, the heads of the twins of Gemini, Pollux, and Castor. Continue clockwise and to the right and down and you'll hit Capella, and then Aldebaran, the red eye of Taurus the bull. Aldebaran leads us back to Rigel, completing the hexagon. Contrast Aldebaran with Betelgeuse, Orion's shoulder, and you'll notice two different shades of red. Compare either with Sirius or Rigel, and the color difference should be obvious. Sirius and Rigel are blue giant stars burning at hotter temperatures than Aldebaran or Betelgeuse. Aldebaran and Betelgeuse are both cooler than the Sun and stars at the end of their lives.

Colored stars have colors because of their surface temperatures. Our Sun has a surface temp of about 6,000 degrees, peaking in the green part of the visible spectrum. Our brains interpret that as being white—containing all colors. Hotter stars will have bluer colors (like Sirius), while cooler stars will have redder colors (like Betelgeuse).

What determines the temperature of the star? The mass. For stars in their normal, hydrogen-fusing lifetimes, the only thing that matters is the mass of the star. Bigger stars are more massive, hotter, and bluer. Smaller stars are less massive, cooler, and red.

Why Are There No Green Stars?

Shouldn't there be? Stars can be infrared, red, orange, yellow, white, blue, and even peaking in the ultraviolet as a sort of ultra-purple. Yet there is not a green star in the sky. But stars have spectra that peak in green. In fact, the Sun puts out more green photons than anything else. But we see light from the Sun as white, not green. This is not an effect of the universe, or stars, but is related to how our brains process light.

The human eye senses color from cells in the back of your eye—three variants of a cell called a cone cell. Depending on how much each cell is stimulated, your brain combines them into a single color. If you perceive light that has roughly equal intensity in all three variants, your brain interprets it as white. For light from the Sun, there is green light but also plenty of blue and red, thus our brains combine them into white light. This is true for any star that puts out a lot of green photons—it will always also make red and blue photons, thus appearing white to our human eyes and brains.

Of course, we can still see green light—it needs to be *only* green photons. If somehow there was a star that *only* created green photons, like a weird giant LED in space, then we would see it as green. But stars don't work like LEDs and instead put out a broad spectrum of light, combining into the beautiful bright white of outdoor light.

Most Stars Are Not Solitary like the Sun

Our solar system is unusual in a few ways, but one of the most important ways is that our Sun is a single star. Many stars are formed not singularly but in pairs or triplets or even quads. Picture Luke Skywalker looking at the sunset on Tatooine in *Star Wars*—gazing at two stars setting beautifully (in a later chapter we will learn if there are planets like Tatooine). We think nearly half of all stars are born with a stellar companion, and the likelihood of being part of a multiple-star system gets higher the higher the mass of the star. The highest-mass stars are nearly always part of a group.

Two stars together are called a binary. They orbit around their shared center of mass, sort of like two people holding hands and spinning. The two stars can be widely separated (thousands of times the distance between the Earth and the Sun, a good unit of measure for solar systems), and take a million years to orbit each other. Or they can be super close and orbit each other quickly, in just under a day.

More than two stars of comparable mass close together becomes complicated—orbital dynamics doesn't always like the "three-body problem." Aside from the sci-fi novel interpretation, the motion of three objects is nearly always unstable. There aren't any closed orbits, every star's motion is different from orbit to orbit (unlikely the highly predictable solar system), and the resulting orbits are chaotic—they can't be defined by their initial conditions.

If three stars form out of the same birth cloud, the most likely outcome is one star gets ejected and the remaining two live out their lives as a two-star binary system (also forming technically the smallest gang of stars possible). The star that gets ejected from the system can sometimes be moving fast, although generally these systems don't have enough energy to create the runaway, super-fast stars that we observe leaving our galaxy. More on them later.

There are some examples of long-lived triple-star systems, but the most stable of these are hierarchal—in that two of the three are a close binary and the third is far out, where the gravity of the two close binary stars acts in unison. The closest stars to our Sun, the Alpha Centauri system, is such a triple system. The closest star to our Sun, Proxima Centauri, is the smaller, more distant third in the system, orbiting the other two at a far-removed distance.

The three-body problem becomes even more of a problem with more stars in the system. A four-body system is inherently even more complex and unstable. The most stars ever observed in a single gravitationally bound system is potentially as many as nine. These systems are rare.

One important note: The instability of the three-body problem applies for objects of similar mass. For something like the solar system, there are way more than just two objects. How does the solar system stay stable? Interestingly, the solar system itself isn't inherently stable—we can only know the orbits for a few million years from now before the complexity of all the parts makes it impossible to calculate. But, most importantly for us, the Sun is the only important object in the solar system. The rest of us, except for maybe Jupiter, are a rounding error. The motion of the Earth cares about the mass of the Sun and cares just slightly about the mass of Jupiter, but nothing else is important. Even for the Sun-Earth-Moon system, the Moon is so close to the Earth that the Earth's mass is the dominant body. If you are interested in more complex orbital calculations, explore LaGrange points and the interesting islands of stability in a three-body orbit that they denote.

My Personal Weirdest Star

We think of stars flying around the galaxy, orbiting, fusing hydrogen into helium, creating light, and generally being amazing. But weirdly, we don't think too much about them joining together. But some weird things can happen in the vastness of space.

A neat type of star is called a blue supergiant (creative name, right?) because they are hot and blue and large. In normal stellar evolution, a blue supergiant is the next phase in the life of a large (ten to 300 solar mass) star after it has fused all the hydrogen in its core into helium. The star begins to fuse heavier elements, and its internal structure changes, causing it to expand outward into a supergiant. Stars should spend little time in this state; it's relatively unstable and the star is just passing through this phase on its way to becoming a red supergiant or, depending on initial mass, a variable giant star. These phases are all the death throes of massive stars.

But weirdly, blue supergiants are overrepresented in the census of stars relative to the short time that the star spends in this incarnation. They should be rare but in fact are observed frequently. There must be some other way of generating this type of star that accounts for the count discrepancy. One theory is blue supergiants can also be made by the merger of two massive stars in a close binary. Most observed blue supergiants are single stars, but most massive stars form as pairs or higher multiples. The theory predicts that a massive star in a close binary with a more ordinary star spiral into each other and eventually merge into a single, weird, massive star. This merged star is long-lived—the merged star will fuse helium into heavier elements. This theory does a good job of predicting the surface abundances of heavier elements in these stars and could be a compelling explanation for the overpopulation of blue supergiants.

While most stellar mergers are explosive (ending in a supernova), these mergers are more like a slow fusion, with the star surfaces kissing, then forming a shape almost like a peanut before eventually coalescing into a sphere. In the huge vastness of space and generally explosive nature of stars, it's nice to know that a gentle, smooth, soft unification is something that happens in the ordinary course of events.

Most Massive Stars

There are massive stars in the galaxy now. Stars of such huge mass are condemned to die in just a few million years, a blink of an eye in the lifetime of the universe. One pedantic note to make is these most massive stars are not the largest stars by radius. The largest stars will be stars at the end stages of their life, which puff up and become large in radius and volume as their mass stays the same or even shrinks.

In theory, stars larger than around 150 solar masses are difficult to make. The process of forming a star is thought to be semi-self-limiting. Once the star is above 150 solar masses, it's hard to get more material to condense onto the star because of the heat and light coming from the core of the star. In addition, it's weirdly tricky to measure the masses of some of the most massive stars. In practice, it's possible there are many stars more massive than 150 solar masses, and we don't know how they are made. That doesn't mean gravity doesn't make them.

A huge fraction of the largest stars we know of (and seven of the top ten) are all in one star cluster called the Tarantula Nebula, which is in the Large Magellanic Cloud, a companion galaxy to our Milky Way. The Tarantula Nebula is a huge, active star-forming region about 160,000 light years from the Sun. The Tarantula Nebula is so active and dense with new stars, and therefore so bright, that if it were as close to us as the Orion Nebula (a star-forming region in the sword of Orion that's roughly 100 times closer to us), it would cast shadows at night. Oh, to live on a planet nearby and see such a thing.

One contender for the largest star in the nearby universe is BAT99-98, in the Tarantula Nebula. Although the mass estimate is uncertain, it could be as large as 225 solar masses. At only seven million years old, it's already at the end of its life and in the end stages for a star of its mass. Its fate is already sealed: BAT99-98 will end up as a spectacular supernova explosion, either leaving behind a black hole or exploding so violently and catastrophically that a black hole can't form. Likely, BAT99-98 will die in the next million years or so, having completed its entire lifecycle in less than the entire history of human evolution.

The largest stars that have ever lived, however, were probably even more massive than the largest stars anywhere in the universe today. At the beginning of the universe, only hydrogen and helium were present, formed in the Big Bang. Without the addition of other elements (like carbon, nitrogen, and oxygen, all formed in the hearts of stars), we think these early first stars formed at higher masses. These are referred to as "population III" stars (continuing a trend of astronomers not being able to name things). We have never directly observed one, but the theory predicts their masses could be up to 1,000 times the mass of the Sun. Such a massive star would have barely had time to form before immediately beginning to die. And its explosive death would spread heavier elements everywhere, helping to enrich later populations of stars. These initial stars would have also created the first massive black holes, potentially creating the seeds for the supermassive black holes found at the heart of every galaxy.

Smallest Stars

If the most massive stars die the quickest, living fiery and explosive lives, then the smallest stars in the universe are the opposite. They are the slowest of slow burns. The smallest stars can live for many times the current age of the universe and will be here long after nearly everything else has burned out.

Technically, for something to be classified as a star, hydrogen fusion must occur in the core (or I guess you can act in a blockbuster movie and become a star, either way). Stars can fuse hydrogen in their cores at about .08 times the mass of the Sun. This is the lower limit of what we would describe as a star. Objects with masses lower than that limit are called either Brown Dwarfs or Jupiter.

Brown Dwarfs are quasi-stars—they aren't massive enough to fuse hydrogen, but they can hold themselves up against gravity by fusing deuterium (also known as heavy hydrogen, or a hydrogen atom with an extra neutron in the nucleus). Deuterium is closer from a nuclear physics standpoint to helium, so the fusion process is easier. Brown Dwarfs are faint, dim, and cool. There is also a brown dwarf limit—the point at which even deuterium fusion cannot happen. In this instance, the object is a gas-giant planet like Jupiter. Jupiter is mostly hydrogen, but it's too small to be a brown dwarf and much too small to be a star. Had things been just slightly different in the birth of our solar system, Jupiter could have been a companion star to our Sun, and the Earth would probably not be here.

Back to the smallest but still technically hydrogen-fusing stars. These stars are so compelling because, while they may not seem as interesting or important as massive stars, they are unstoppable powerhouses. A 0.1 solar mass star will live for so long that it feels infinite. The Sun will live for ten billion years, but a 0.1 solar mass star will live for ten trillion years, 1,000 times as long as the Sun, already living for a long time. Ten trillion years is difficult to conceive—nearly 1,000 times longer than the current age of the universe. Whole galaxies will form and merge and recondense and merge again and exhaust all of their star-forming hydrogen, and a 0.1 solar mass star will not yet have reached middle age. The Sun will have died, leaving behind material to form the next generation of stars, and generations after that, and even more generations, and a 0.1 solar mass star will have barely noticed the passage of time. What, to a 0.1 solar mass star, is the depth of infinity?

Why do these stars live so long? Essentially, it's because they are efficient. These stars turn nearly every atom of hydrogen in them into helium. The Sun, in contrast, is only about 10 percent efficient. Most of the hydrogen in the Sun will never become helium. The Sun isn't perfectly mixed; it's more like a layered onion. Only the hydrogen in the center gets converted into helium. The rest is stuck in the cooler outer regions and isn't accessible for fusion. But a smaller star is more convective—the whole star is in a continuous churn like a pot of boiling water. This causes the star to be well-mixed, and while the core is fusing hydrogen into helium, fresh hydrogen is being brought in from the outer regions of the star. Thus, a 0.1 solar mass star is nearly perfectly efficient, fusing as much hydrogen as the Sun overall, despite being one-tenth of the mass. In addition, the rate at which hydrogen is fused is super slow—these stars don't need to be hot, just enough to counteract gravity, and so they fuse at a slower rate than a star like the Sun. Just one-thousandth of the fusion per second is happening. These two things—efficient use of hydrogen fuel and slower burn rates—give small stars their incredible lifetimes.

Fastest Stars

Keeping track of all the stars in the galaxy is a complex task. Not only are there billions of them, but they all are always moving in a complex ballet directed by gravity. Some, like the Sun, float around the galaxy mostly in the plane (our galaxy is flat like a pancake), orbiting around the center of the galaxy over millions of years and generally ordinary (FYI, a galactic year is 225 million Earth years). Other stars take a more exotic path.

There is a class of stars that, even for speeds in space, are traveling in a real hurry. These stars are referred to as "hypervelocity stars," for obvious reasons, and generally have speeds above ten times the usual galactic average (this corresponds to roughly 1,000 km/s). We think there may be about 1,000 of these stars (out of 100 billion), although we've only observed twenty of them. There are two possible ways to get a star moving this fast.

The first way is via a close encounter with the monster under the bed of our galaxy—the supermassive black hole at the center. A binary star system (which we learned earlier) could have a gravitational close encounter with the supermassive black hole in which one star gets captured by the black hole and goes into orbit around it (without being eaten). In this scenario, the energy of the binary system gets converted into an ejection speed for the other star in the pair. One gets captured and the other is flung out at incredible speeds.

This mechanism only works for hypervelocity stars leaving the galaxy, which, if you trace their trajectories backward, point toward the galactic center and its lurking black hole. For some hypervelocity stars, this isn't the case. Instead, they have also started life as binary stars, but one star in the pair explodes in a supernova and the other pair is blown off course, ingesting a huge amount of energy in the form of a velocity kick. The newly single star can be traveling so fast that eventually it will escape the gravitational pull of our galaxy, ending up in the vast emptiness of intergalactic space.

What this means for us: There are many other superlative stars—stars that, even in a sky full of stars, in a galaxy with as many stars as grains of sand on a beach—that are unique and unusual. Stars that flare, stars being vampirically consumed by a companion, stars spinning super quickly. But none are so unusual as the one around which we revolve. The Sun itself was once thought to be an ordinary star, but as we learn more about stars, we find that it is rare and special. It's larger than average but in the boring outskirts of a boring galaxy. It's quiet, as far as stars go; it's spinning, but not too quickly or too slowly, and has maybe an average number of planets, although they are strangely arranged. It's the precise best place to be a human that we know of, which makes it superlative indeed.

CHAPTER 6

Nuclear Spaghetti and the Deaths of Stars

Weird Facts: Stars are born to die, like us all. The death of a star can be extravagant—some are so bright they can be seen across the universe. Stellar death is also an opportunity to explore some of the weirder forms of matter. As we will see, remnants of massive stars sometimes are full of spaghetti. Read on to learn more.

No star can shine forever. While stars are *the* important feature of our universe, the universe's response to gravity means they all have an endpoint. For each star, this is defined by the point where it can no longer continue to fuse hydrogen into helium, either because it has used it all (in the case of small stars) or has used all available hydrogen in the core of the star (in the case of large stars). Without that readily available fuel source to continue fusion, the star can't resist the crushing pull of gravity, and something gives way. The exact sequence of events that lead to a star's death is similarly dependent on what we saw last chapter—it all depends on the mass of the star. As in life, so in death.

The Mass of the Star Is All That Matters

We learned that the only determining factor in the type and lifespan of a star is how massive it is. This is largely also true for the old age and deaths of stars. During their hydrogen-fusing lifetimes, most stars are not changing. They may get a few percent brighter over the course of billions of years. Their rotation may change slightly. They may become more or less active. But generally, their time burning hydrogen is notable only for its incredible stability. The star acts as its own self-regulating thermostat—if, let's say, the core of the star got slightly hotter, it would fuse more rapidly, generating more outward pressure against gravity. With more outward pressure, the core will get bigger, and because of thermodynamics, cool. The fusion rate decreases, the pressure gets a little less, and the core shrinks again. Generally, the star settles down at a nice equilibrium—a state where all forces are balanced and, if you are a creature living in the glow of said balanced star, you enjoy the sunshine without thinking too much about it.

But all stars must die. The fate of a star can be split into roughly three categories—small and moderate-sized stars that effectively burn out, massive stars that explode and leave behind a remnant, and massive stars that explode and create black holes. The fate of the star on one of these three paths is essentially already chosen when it is born. All that matters is the mass.

The Death of the Nearest Star

In terms of stellar deaths, the one we should be most concerned about is that of our precious Sun. The Sun, being an average-sized star, will not die in a spectacular explosion. Instead, it'll slowly burn out, ending its beautiful life as a fading ember, but not before destroying the Earth. Of course, life on Earth will be long dead before the end of the Sun—the Sun is getting brighter by about 1 percent every 100 million years. In about 600 million years, it'll be bright enough and hot enough that Earth will no longer be in the habitable zone and the surface will be too hot to sustain life. Eventually the oceans will boil away and our once verdant pale blue dot will be nothing but a desolate rock.

But it gets better. After the death of the Sun, even the desolate rock that Earth turns into will be long gone. After the Sun has exhausted the hydrogen in its core (it will only be able to access about 10 percent of the available fuel in the whole star), it will need to switch to a different fuel source. Effectively that means fusion briefly pauses because of no available hydrogen. This causes the outward pressure holding the star up to decrease and gravity briefly wins against pressure and the Sun will start to shrink, getting hotter and brighter. As the Sun contracts, it will heat up and reignite hydrogen fusion in a thin shell just outside of its core. This shell-burning also has the effect of puffing the star to a large size and cooling the outer layers, creating an intermediate type of star called a "red giant" due to its size (giant) and color (red; once again, a creative name, right?). About a billion years of time will pass, and the core of the Sun will continue to shrink until it reaches a high enough density and temperature that it can start to fuse helium into carbon. This event happens simultaneously across the core (it's called the "helium flash") and then the Sun will settle down again for about 100 million years of hydrogen shell fusion and core helium fusion.

Of course, this settling down will happen with a radius about 250 times larger than its current size, easily engulfing Mercury, Venus, and likely Earth. Even if humanity somehow made it another 4.5 billion years (on an Earth that can no longer sustain life, either because of our actions or the eventual action of the Sun), the people left behind will have a magnificent and awful view of the Sun slowly enlarging on the sky—as inevitable as a tsunami of fire approaching them with scorching power. Long before it breaches the atmosphere, all possible observers will be dead with no one left to report on what's happening. Hopefully for us, we will have taken to the stars billions of years before, finding the warmth of other suns and new worlds to live on, watching the destruction of our home world from the safety of another.

After this red-giant phase, the Sun will enter a period of great instability—alternating between running out of fuel, shrinking, reigniting in successive shells of hydrogen and helium burning, stabilizing, and then running out of fuel again and restarting the cycle. These stages also include significant mass loss from the Sun; its outer layers become only loosely tethered to the star and eventually get launched out into space. This mass-loss process, which happens to many smaller stars at the ends of their lives, creates some of the most beautiful, ephemeral nebulae in the galaxy. Rings and shells of bright gas glowing in the darkness. These nebulae last only for a few thousand years, just a blink of an eye in cosmic time.

The dense, hot core of our Sun will live nearly forever though, continuing as a "white dwarf." A white dwarf is a dead star made of the hot embers of a former sun. After losing its outer layers, the core of the Sun remains, consisting by now of mostly carbon and oxygen and is just half the mass of the original star. The solar white dwarf is tiny—only about the size of the Earth—but incredibly dense and hot (around 10^8 degrees C, much hotter than the surface of the Sun is now). The white dwarf still has to contend with the pull of gravity but is slowed from further collapse by "electron degeneracy pressure." More on this later.

But the core will last forever—slowly cooling until it becomes so cold it's recharacterized as a "black dwarf." The eventual fate of the Sun's white dwarf remnant is likely to slowly cool over trillions and trillions of years, becoming, along with other white dwarfs, one of the last remaining dense objects in the universe. Depending on if protons decay (a hugely important question that we don't know the answer to), the now-black dwarf may live on _electrons + be packed together infinitely dense due to Pauli exclusion principle. As a star collapses & its density ↑ electrons are forced_

for anywhere from 10^{20} years to 10^{200} years, an unfathomable amount of time. So, while the Sun's hydrogen-fusing life is over in a blink, its corpse will roam the universe for eternity.

to occupy higher energy levels, creating

Making Degenerate Matter

outward pressure that counteracts gravity

Stars are held against gravity because they are fusing hydrogen into helium, creating heat and energy and an outward pressure that keeps the star from collapsing. But the dead corpses of stars don't have active fusion occurring. There is no outward, fusion-driven pressure against gravity. And they don't collapse into black holes. The forces holding the dense, dead cores of stars are much stranger and more unusual; we refer to these forces as degeneracy pressure. This is likely a term you've never heard before—it relies on the quantum mechanical properties of electrons and nuclear particles.

In the case of a white dwarf, the star is held by a quantum mechanical property called the Pauli Exclusion Principle. This principle says that electrons can't occupy the same states around a nucleus—effectively you can't pack more than a certain number of electrons into a small volume. As the core of the star gets denser and denser, it eventually becomes supported by this principle— more electrons can't get packed into such a small volume and the electrons exert an outward force. This electron degeneracy pressure is what ultimately counteracts the pull of gravity for the dead core of the star. This pressure can only hold white dwarfs to a certain mass—1.4 times the mass of the Sun. If the white dwarf somehow becomes more massive than that limit (for example, because it has a companion star that it is stealing mass from, a common event in the universe), the pressure of electrons can't hold it up anymore and it will collapse. This results in a spectacular supernova explosion that destroys the white dwarf.

The white dwarf that our Sun will turn into won't end in a supernova explosion, however. There is no companion star for it to siphon material from, so it'll remain as a roughly 0.5 solar mass white dwarf, cooling for eternity.

Nuclear Spaghetti

Another type of degeneracy pressure runs the show for the cores of intermediate-mass stars—about eight to twenty-five times the mass of the Sun. These stars have shorter lifetimes and die in spectacular supernova explosions. While the Sun enters successive states of shell fusion and helium fusion, more massive stars go through many different fusion processes in the core before they die. These stars fuse hydrogen into helium, helium into carbon, carbon into oxygen, oxygen into silicon, and silicon into iron. Iron is the end of the road for the star—fusion of iron into heavier elements doesn't create energy the way that fusion of elements lighter than iron does. Instead, fusing iron uses more energy than it makes, and the star can't continue fusion. A core of iron builds up that isn't being held by any outward pressure and, all at once, the star cannot support itself. The star effectively collapses in less than a second, crunching the core to a tremendous density. The outer layers of the star rebound outward in a spectacular supernova explosion, leaving behind the ultradense core as a remnant. For massive stars, a black hole will form, but for intermediate-mass stars, the remnant is a neutron star.

Neutron stars are objects so compressed that the electrons and protons in the star have merged into neutrons. The entire star is effectively an enormous solar-mass-sized nucleus. Aside from black holes, these are the densest objects in the universe. They are so dense that we don't fully understand the physics of their interiors. Neutron stars are held up by neutron degeneracy pressure, which is like what holds up white dwarfs except with neutrons instead of electrons. This follows the same exclusion principle as before—two neutrons can't occupy the same space and thus there is an outward force when you cram them too close together. Neutrons can be squished effectively, which makes a solar-mass neutron star tiny—a sphere just ten to twenty kilometers in diameter. And they can be as massive, or more massive, than the entire Sun. Truly weird objects.

One open question in physics is what the state of the matter is within a neutron star—the pressure exerted on the neutrons gets higher and higher the deeper into the star you go. One theory is the existence of "nuclear pasta" in the first 100 meters under the surface of the neutron star. Under changing pressure conditions, the neutrons will organize themselves into semi-stable configurations, most of which resemble types of pasta. Nuclear spaghetti, clearly the best term for any structure in the universe, could occur at intermediate depths into the neutron star, although there is also a gnocchi phase at the upper levels, and a bucatini phase deeper down. Eventually the pressure is so great that you can't even make pasta out of the neutrons anymore and you end up with unusual states of matter that we only have theoretical calculations for.

Neutron stars are weird, hugely unusual objects. Their discovery by Jocelyn Bell Burnell in 1967 helped astronomy to realize degenerate objects (white dwarfs and neutron stars) were real and black holes could be real too. Jocelyn Bell Burnell observed a regular repeating radio pulse from space, which she originally identified as "LGM-1"—little green men. But it was a type of rapidly rotating neutron star that emits a radio pulse from its poles as it spins, like a lighthouse. It just so happens that the beam of the lighthouse is pointing toward us during each rotation. We've sense found many of these "pulsar"-type neutron stars. We've also found neutron stars with strong magnetic fields called "magnetars." The fields are so strong that if you approached one, all the electrons would be ripped out of your body, and you'd die almost instantly. Not a good way to go.

Neutron stars, like white dwarfs, will likely live nearly forever. Slowly cooling over trillions and trillions of years. Floating around the universe long after we are all gone. More on that later.

What this means for us: Ultimately, the deaths of stars are all part of the normal lifecycle of a galaxy. Stars die, creating new elements and conditions that let new stars form. The star is a relatively simple object—mostly made of one thing and mostly governed by gravity. But even within these simple constraints, stars are incredibly varied, and their deaths are just as interesting. The dead remnants of stars will remain much longer than anything else, outlasting even the longest-lived stars.

CHAPTER 7

We Used to Think This Was All There Was

Weird Facts: Every human culture has stories about the stars. Every single one. Making up stories about the lights in the night sky, wondering about their origin, is something that all humans have done for all of time. When looking up at the night sky, we are all the same.

Let us take a step back here. We have been considering the weirdness of spacetime, the singularity of a black hole, and the fundamental unit of the universe, the star. Now let's re-examine our place in the universe and how we figured out that place. The term *cosmology* is used to describe any account or theory of the origin of the universe. Later in this book, we will describe our current best understanding of the origin of the universe, the Big Bang. Early humans came up with their own cosmologies to explain the origins of the world around them. Those too are cosmologies. We've spent thousands of years moving from cosmologies informed by myths to cosmology informed by facts and data. In making those cosmologies, we are always held back by our desires and limitations.

In the history of science (maybe in all of history), one constant throughline has been a certain level of self-absorption on the part of humanity. Everyone thinks they are the center of the universe (all of those people have been wrong, of course; in fact, *you* are the center of the universe). When humanity has tried to develop a concept of where we exist in the universe and how we got here, the default has always been to place us at its center. We are the main characters in the story and the ones who create action through our existence. Early cosmologies, the history of the universe and creation, all focus on how the Earth and its land were formed and how we came to be here. In some sense, our modern cosmology is still concerned with this but sees (maybe somewhat more accurately) our presence in the universe is incidental to the larger story.

Let's start with early tales of the formation of the universe. Supernatural beings always bring about creation—gods or deities who represent some natural phenomenon. Generally, the myth has a story structure, and humans and creatures can speak or be transformed. They all take place in an unspecified but distant past. While there are over 100 different versions of creation across human cultures, they fall into a few categories. In the Bible, the Christian God brings about order out of chaos, creating light, then day and night, then eventually the Earth and all its creatures. The exact mechanism of creation isn't described and it's not clear if there was nothing before or if God sculpted the world out of some preexisting material. In the Finnish creation myth, the Earth forms out of an egg that splits open. In the Iroquois Native American creation stories, there was only water and a spirit realm far above the water. Water animals dive below the water to eventually bring up mud that grows to form the land, creating Turtle Island. In Chinese creation stories (of which there are many) there is no agent, no God or deity or cause doing anything. Creation happens; order emerges out of chaos. In ancient Greek creation, the first gods emerge out of nothingness, and give birth to later ones, eventually forming the full pantheon and all creatures.

A commonality of creation stories is the human desire to explain our origins. Each story gives specific details on the origin of humans, who made the first humans, and sometimes why, providing motivation. For these myth-based cosmologies, the primary concern isn't with explaining the motions of things in the sky, or the presence of night and day, but more with explaining why humans are now out in the world, doing all sorts of wacky things. Sometimes the lights in the sky and their motion are explained, sometimes not. At the least, usually the Sun and the Moon play a key role in the story.

Over time, we have moved from a myth-based cosmology to one that attempts to explain the world around us mechanistically, starting with the ancient Greeks. They had an origin story with gods and goddesses, forming things and ordering the world. But ancient Greek philosophers moved beyond the myth into something unusual for their time. The innovation of Greek cosmology was it removed the supernatural from the story and attempted to derive a system of the world that was entirely mechanistic and self-contained. They aren't explaining how the world formed, but how it continues to move. The Greeks didn't need to explain why, but they attempted to order the mechanisms that drove their observations of the world around them. This didn't mean they overthrew the concept of having gods, but rather that a god's intervention wasn't required to keep the world turning. Apollo doesn't need to wake up and drive the chariot every morning to ensure the Sun travels across the sky. The Sun just does it. The world works naturally on its own.

The first step to develop this system is to move from a flat Earth supported by a structure (sometimes pillars, sometimes a turtle or many turtles, sometimes an ocean) to an Earth (of any shape) that exists but is not supported by anything. Anaximander of Miletus (610–546 BC) was the first person to record their theories of other ways for the Earth to be. He described an Earth that floats in infinity, unsupported by anything but remaining in the same location by "indifference." Anaximander believed the universe was ruled by laws, just like human societies, and understanding the universe was a question of figuring out the laws. His concept of the Earth was still flat; the floating Earth was a squat cylinder shape, and all of Greek civilization was on the flat top. This idea that the Earth could float in nothingness was a huge leap forward, which allowed for all sorts of other innovations. If there is nothing below the Earth, then the paths of stars and planets around the Earth can be easily explained. Stars that disappear over the horizon to the West and reappear on the eastern horizon hours later could have traveled underneath the Earth. Simple as that.

Once you allow the Earth to float in a void, observations of the motions of everything in the sky fall into place quickly. Seven objects move across the sky in slightly different ways—the Greeks referred to these objects as the seven planets. The Sun (which they considered a planet) moves across the sky on a regular path every day. The Moon (also a planet to them) moves across the sky regularly but on a different schedule and trajectory than the Sun. Mercury, Venus, Mars, Jupiter, and Saturn each move in their own times and paths. Against all of this activity lies the backdrop of the stars, which move as a unit. The Greeks interpreted these motions as a series of spheres; each of the seven planets had its own sphere, nested within each other, and the celestial sphere that contained all the distance stars was the outermost and slowest. This view was solidified by Aristotle as the perfect concept of the universe.

The Aristotelian concept of the order of the universe places the Earth at its center, with everything (Sun, Moon, planets, stars) orbiting around it. The Earth was the unmoving, stationary pivot around which all of reality turned. This is just as hubristic as the most foolish Greek hero thinking they are better than the gods. The absolute gall of the Greeks to assume that the Earth was the pivot around which the entire cosmos revolved. So self-involved.

But can we blame them for this? If you don't know anything about orbits or distances or sizes or what any planet is made of, it's almost natural to assume that the Earth must be the center. Standing outside and watching how things move across the sky, it appears that everything rotates around the Earth while the Earth itself remains stationary. The Sun appears to rotate around the Earth. The Moon actually does rotate around the Earth. The motions of the stars and other planets also appear to rotate around the Earth.

This is just an aftereffect of perspective—Einstein tells us that motion is relative to the position and state of the observer. If we were living on the surface of the Sun, it would be obvious to us that everything rotates around the Sun. If we lived on Jupiter, it would be just as obvious that everything rotates around Jupiter. The complexity that took many hundreds of years to figure out is that Earth's rotation on its axis contributes to some but not all of the motion we observe. Some of the motion is from the Moon rotating around the Earth. Some of the motion is from the Earth spinning on its axis. And some of the motion is from the Earth (and other planets) rotating around the Sun. Three types of rotation are happening at the same time. This mixing of modes may have been why figuring out our place in the universe was so hard and took thousands of years.

If the Earth didn't have a Moon, might we have considered other arrangements earlier? It's possible that we would have dug more into other geometries (aside from the Earth as the center of everything) more seriously, considering the errors and flaws in the Earth-centered system. The fact that the Moon does orbit the Earth makes it harder to leap intellectually to the possibility of other motions, to consider if the motion of the other planets and stars can be explained by other arrangements. Aristarchus of Samos, another ancient Greek astronomer, attempted to calculate distances to the Sun and Moon and estimate their masses. He proposed that maybe the stars are similar in makeup and character to the Sun but only appear different because they are far away. He also suggested that maybe the Earth rotates around the Sun. Since the Sun seems to be the biggest object around, and is much bigger than the Earth from his calculations, shouldn't it be at the center? But Aristarchus lacked a way to develop strong evidence for his theories and for 1,000 years, the Earth as the fixed center of everything was the dogma of the Western world.

The Greek astronomers got a lot of things right, though—they figured out that the Earth was a sphere and roughly estimated its size. They predicted (with some small errors) the motions of planets of the sky. They figured out the tilt of the Earth, and that the Moon was much closer than the Sun. But the true scale of the universe was unknown. The distance to the "fixed stars" could never be determined. The idea that the distance to another star might be almost incomprehensibly vast was an outlier (still proposed by Aristarchus anyway) but not taken seriously.

Before moving on from the ancient world, it's useful to explore the ancient Chinese perspective on this. Chinese astronomers kept meticulous records for thousands of years, predating ancient Greek astronomy. These records give us evidence of supernova explosions in the recent past and were accurate enough to predict eclipses. Indeed, ancient Chinese astronomer's entire motivation was to predict the motions of the Sun and Moon, and they were less concerned with other planets or understanding why the motions happened. The Chinese universe was less structured than the Aristotelian one. One model had a celestial sphere of stars with the Earth at its center, but other models were infinite, with the Earth moving but not in any way that was noticeable to humans.

And for the most part, this system of the world was perfectly adequate. Predicting the motions of the planets and stars was important for predicting seasons and calendars. Certain cultures cared about this more than anything else. In ancient China, imperial astronomers were tasked with making predictions of Moon phases and eclipses and their ability to be precise and correct or incorrect could end dynasties. If they predicted the phases of the Moon wrong, their life was on the line. In other societies, the motions of stars and planets were important for astrology and trying to make predictions about the future. So, positions were important but not understanding why. And a flawed system that made decent predictions was fine for your normal medieval astrologer.

One step forward happened in the early 1500s. Polish astronomer Nicolaus Copernicus, after a lifetime of observation and study, published a book that started a revolution (it is why the term *revolution* means both to orbit around something and to overthrow a system). *On the Revolutions of the Heavenly Spheres* suggests the Earth isn't the stationary center of the universe, as was assumed for 1,000 years, but the Sun is the object around which all things orbit. Copernicus was inspired in part by the work of astronomers from the medieval Arab world, scholars and polymaths working in modern-day Iran, Baghdad, Syria, and Al-Andalus, the Islamic-ruled region of Spain. These Arab astronomers argued as early as the year 1000 AD that the Earth should rotate on its axis and proposed spheres centered around the Sun as an ordering of the universe. Like all scientists, Copernicus's work was set in a larger context of those who came before him. Copernicus's publication unleashed a scientific revolution in Europe, inspiring later observers like Tycho Brahe and Galileo Galilee and bringing down the wrath of the Catholic Church, which believed the Earth was stationary.

But Copernicus's work was still just a theory, slightly more advanced than the Arab astronomers who inspired him and more advanced than Aristarchus of Samos who first thought of it. But no evidence was available to prove which theory, heliocentric (the Sun at the center) or geocentric (the Earth at the center), was right. Our ability to see the real arrangement of the universe didn't start moving forward until the invention of the telescope in the early 1600s. This innovation meant that it was finally possible to see celestial objects up close to determine if the idea of a perfect heavenly sphere was accurate. Galileo (who didn't invent the telescope but improved it) was the first person to point it at the sky and publish his results. He used this new technology to observe objects in our solar system and found them to be unusual.

The Moon had craters, and its surface was pocked and scarred, not a perfect beautiful Aristotelian sphere. Jupiter had a red spot on it and four "stars" that appeared to follow it around. Galileo realized these were moons of Jupiter, orbiting around the planet just as our Moon orbits around ours. Galileo observed Venus, discovering that it has phases, just like the Moon. He observed the Sun, discovering sunspots. Galileo worked in a time when new ideas about the ordering of the universe were being proposed. His observations convinced him that Copernicus's work from decades earlier was right—the Sun was the center of everything. Galileo famously got into a great deal of trouble with the Inquisition in Rome and ended up living the last years of his life under house arrest. It may be of some consolation to him that, of course, he was right about a lot of things and received a posthumous apology from the Catholic Church in 1992.

Galileo's contemporary Johannes Kepler proposed a more precise model of the solar system, in which the Sun was at the center and the Earth and other planets orbited the Sun in ellipses, not perfect circles. The Keplerian model of the solar system perfectly predicted the orbits of the planets (well, except for Mercury, which wouldn't be explained until general relativity). Kepler's model helped inspire Newton to develop a theory of gravity that was used for 400 years until it was overthrown by Einstein's general relativity. General relativity will hang around for a while until we find an even more accurate description of the nature of gravity. And so on, as every theory builds on the last and gets a little close to the truth before being overthrown by a better theory. Ad infinitum.

As an aside to the improvements made in our theory of the universe, work by scientists in this time moved all of science forward. The invention and use of the telescope started a revolution in astronomy and, in part due to the meticulous work of Galileo in other non-astronomy fields of science, helped to usher in the use of the scientific method. The scientific method is the reason we know anything about our *Weird Universe* to even explain in this book. It's the reason why we have cell phones and computers and airplanes and laparoscopic surgery. The scientific method is a concept that allows one to explore the world and universe in a methodical, systematic, and careful way, creating, then refining and testing ideas to continuously hone a model of the way the world works, to understand how anything functions. The scientific method consists of a cycle of developing a hypothesis (a question that will be proved right or wrong) or theory to test, conducting an experiment or observation that tests the hypothesis, determining if the outcome of the experiment matches the expectation, and starting the cycle again with more information. The model is never the *truth* or complete picture of reality but is an approximation that gets more accurate with every cycle. In understanding the ordering of our solar system, a chain of people over thousands of years all contributed ideas and suggestions. In a similar chain through the ages, the scientific method slowly grew. Eventually it was also Galileo, when he wasn't getting in trouble with the authorities or staying up all night, who added in two key components—the use of experimentation and the addition of mathematical calculations. Experiment and theory are the cornerstones of the modern scientific method.

Yet, even in this more accurate view of the universe, one with the Sun at the center, huge questions abound. Why is the Sun at the center of the universe? How big is the universe? How far are the distant stars?

To answer these questions, better observations are needed. Only with the invention of the photographic plate does astronomy take its next leap forward in understanding. The distance to stars wasn't known in any way until 1832, when the first measurements to a handful of nearby stars were made after about 200 years of improving telescope technology. Those measurements revealed the stars were unimaginably far from us—distances measured in tens and hundreds of trillions of miles, at such distances that it made the solar system look tiny. Our view of the size of the universe expanded.

This trend continues—we think that the universe is small and moderately sized, but it turns out we are (of course) wrong. With better and better telescopes, astronomers discovered more extended, diffused objects in the night sky. Some of these have been known since time immemorial, called Nebula, from the Latin word meaning "mist" or "cloud." Ancient astronomers from Greece and the Middle East noted stars with halos or clouds around them. The Great Nebula in Orion (also known as the sword hanging from Orion's belt) is a fluffy cloud in the sky, visible to the naked eye but even more interesting when viewed with a telescope. The Great Nebula of Andromeda is also a fluffy cloud that has always been visible to the naked eye but only came into sharper relief with photography.

The Great Nebula of Andromeda appeared to astronomers to have a spiral structure and there were debates for hundreds of years about its distance. The question was this: Was Andromeda a small, nearby nebulosity may be in the process of forming a solar system (and thus the spiral shape), or was it an "island universe," a concept proposed by Immanuel Kant, which was farther away and larger?

Once again, the assumption was our observable universe was all that there is. This framework seeks to fit anything that we observe into the context of things we already understand, in this instance solar systems and nearby, easier-to-observe nebulae (the plural of nebula) and clouds. We shouldn't be too hard on people in the past for this behavior. It takes a great leap of imagination to come up with a concept that doesn't exist. The debate about the true nature of Andromeda and the larger question of whether anything was outside of our own Milky Way galaxy was definitively put to rest in the 1920s.

Edwin Hubble measured the distance to Andromeda using multiple images taken over time, relying on calculations from Henrietta Leavitt, who did groundbreaking work on the predictable variability of certain stars. Henrietta Leavitt was a Harvard astronomer and the first person to figure out a reliable way to measure the vast distances between stars and galaxies. While she never earned a PhD (due both to chronic illness and the common discrimination against women in higher education), she worked at the Harvard College Observatory, examining photographic plates—old-fashioned images of the sky taken on sheets of glass. This work was tedious and required incredible attention to detail. Leavitt discovered a star known as a Cepheid (named after the first of its kind to be discovered, Delta Cephei), which pulsated predictably. These stars are in the end stages of their lives and have a variability—they get smaller and hotter and then larger and cooler and then back to smaller and hotter and so on, almost as if the star is breathing in and out. These types of pulsations are related to an instability in how photons from the star's core are captured by the outer layers of the star. Leavitt first noticed that there was a relationship between the length of the pulsations—how long the inhales and exhales of the star took—and how intrinsically bright the star was. Shorter period pulsations were connected to intrinsically fainter stars. If you know how bright a star is, it's easy enough to calculate how far the star is. Thus, a distance for something that was previously difficult to figure out.

Back to Hubble, making measurements at Mount Wilson, outside of Los Angeles, he found a Cepheid variable star in his observations of Andromeda and used it and Leavitt's work to calculate a real distance to the "Nebula." His measurement indicated that Andromeda was vastly farther away from us and must be its own "island universe" or galaxy, separate from the Milky Way galaxy. His original calculation was nearly one million light years. He was off by a factor of two—the true distance is a bit over two million light years—because of subtleties in the variable star he was using. In any case, his discovery, building on hundreds of years of discovery, proved that the universe was much larger and more interesting than one that is only the size of our small Galaxy. His discovery meant that not just Andromeda but every observed galaxy went from being a tiny part of the Milky Way to a part of a vast and distant cosmos. Once again, the size of the universe gets larger.

Even today, we are limited by the cosmic horizon—the distance that light has traveled since the start of time. What is beyond that? Is it more of the same or something more unusual and interesting? Let's hope the universe continues to surprise us and something weird and wonderful awaits just over the horizon.

What this means for us: In a few hundred years, we've gone from a tiny clockwork universe with the Earth unmoving at the center to one that spans billions of light years with massive spirals of stars and furious fusion powering points of light that travel across ages to reach us. As we've discovered more, and as we shall see in the coming chapters, we've realized that the universe still holds mysteries for us to explore. There are countless unanswered questions for discovery. And despite learning this lesson over and over, we have never gotten over ourselves, always assuming what we see and have already experienced is all that there ever could be. Many discoveries in astronomy and science have been made by people who have the enviable ability to look beyond what already is and imagine something different and un-Earthly in the most literal sense.

CHAPTER 8

Up Until 1960, We Thought Venus Was a Tropical Paradise

Weird Facts: Scientists were convinced that Venus, the second planet from the Sun, was a tropical paradise up until the 1960s. Venus has a thick, cloudy atmosphere, and the assumption was that clouds on Earth mean rain and forests, so surely, they mean the same thing on Venus. But in fact, the clouds of Venus are made of sulfuric acid (a super dangerous and corrosive substance), in an atmosphere almost entirely composed of carbon dioxide, and a surface temperature of 870 degrees Fahrenheit. Venus is barren, parched, boiling, and decidedly not a paradise.

In our human-centered view of the universe, as we learned in the last chapter, no place has been subject to our prejudices as much as the solar system. Aside from our false certainty for hundreds of years that the Earth was its center, we've made a lot of specific assumptions about the planets and how closely they resemble Earth.

In our exploration of the solar system, we've learned a lot about what goes into making a planet (rocky versus not), how important the Sun is for the climate on every planet (varying from important to meh), and how unusual our place in the solar system is. Of course, that knowledge has come through hundreds of years of observation, decades of physical exploration, and some jarring mistakes in our understanding along the way.

Historically, humans have known about five planets in the solar system since there were first humans. The five closest planets are so bright in the night sky that they are easily seen by the naked eye. Aside from their brightness, they travel across the sky on different paths than the regular motion of the stars, as well as from each other, the Sun, and the Moon. The word *planet* comes to us from the Ancient Greek word for "wanderer"—the planets were called *asters planetai*, wandering stars, derived from the verb "to wander," *planasthai*. These wanderers included the Sun and the Moon, and the ancients understood there to be seven planets— part of the reason we have seven days in a week. In some languages, every day of the week is named after a celestial body. In English, the remnant of this is Sunday for the Sun, Monday for the Moon, Saturday for Saturn. Tuesday to Friday are named for Norse gods, but in many other languages, Mars, Mercury, Jupiter, and Venus reign supreme once a week.

Interestingly, the naming convention is out of order from the ancient understanding of the distance of the planets from Earth (which at the time was, from slowest to fastest moving across the sky: Saturn, Jupiter, Mars, the Sun, Venus, Mercury, the Moon). This is assuming all these planets were orbiting around the Earth (wrong, as we know) but isn't so bad as far as an ordering goes. The days of the week order is based on an ancient astrological system called the Chaldean order. The day of the week is named after the planet, which "rules the first hour" of the day in this system. There is some trickery to how this is ordered, and thus we end up with our current convention (Sun, Moon, Mars, Mercury, Jupiter, Venus, and Saturn). These seven days and their names have lasted millennia and will probably continue to do so as long as there are humans.

Similar to our long knowledge of these planets, humans have always been aware of the existence of comets and meteors. These are bright enough that they are easily noticed by even the most incurious stargazer. One of the earliest recorded depictions of a comet was Halley's Comet, which is embroidered into the Bayeux Tapestry, depicting the Norman conquest of England in 1066. The comet appears as a star with a tail, and the Latin words "isti mirant stella," meaning "they marvel at the star." Everyone in that tapestry panel looks upset. And for good reason: comets were a harbinger of catastrophe and ill omen. Rulers fell and kings died when a comet flew by. Ancient China kept meticulous records of the comings and goings of comets for nearly 3,000 years. Aristotle believed comets were caused by the same mechanism that made the aurora and meteors—activity in the upper atmosphere. As always with Aristotle, he got a little right and a lot wrong, but he was so convincing that his view lasted for over 1,000 years. One part that he got right is that comets leave behind a trail of pieces of themselves that can sometimes hit the Earth's atmosphere and create meteors. All of our favorite annual meteor showers are comet tails falling to Earth.

We now know that comets and meteors are both remnants of the early solar system formation. Comets are small, icy balls of frozen slush that usually reside far in the distant edges of the solar system in two regions: the Kuiper Belt, where Pluto also resides, and the Oort Cloud. Comets are usually happy to hang out far from the Sun, frozen and tranquil. But occasionally their orbits get perturbed and begin the long journey toward the Sun. The tail of the comet is generated when it gets close enough to the Sun that the outer ice layer of the comet is melted and streams behind it.

In contrast, a meteor is any object entering the Earth's atmosphere and causing a streak of light, like a shooting star. Meteors can have many sources—some can be from comet fragments, some can be from bits of asteroids, and some can even be broken pieces of other bodies (like the Moon or Mars).

Ancient people noticed the presence of both planets and comets because they represented changes in the normal order of things. After the invention of the telescope in the early 1600s, we found a few more planets—Uranus in 1781 and Neptune in 1846. For many hundreds of years, starting in the late 1500s after Kepler measured the distance to the Sun and orbits of all the known planets (at the time, five planets), astronomers searched for a planet between Mars and Jupiter. There was a clear gap in what is otherwise a well-ordered pattern in the distance between planets. This beautiful ordering is that every planet is about twice as far from the Sun as the planet before. Venus is twice as far as Mercury, the Earth is twice as far as Venus, and so on. This pattern is likely an effect of orbital resonances—the gravitational presence of a large planet (like Jupiter) will create regions of space around a star that don't have stable orbits and thus, will not have a planet. All planets do this to a greater or lesser degree—clearing out the orbits near them so they are the only ones around.

It's obvious after seeing the pattern that there is a missing planet between Mars and Jupiter. Jupiter is about four times as far from the Sun as Mars is, not double. As always with humanity, it's hard to imagine something you've never encountered before, and thus astronomers searched for a planet in this area because they couldn't think of what might be there instead. Eventually they found a few small fast-moving dots at the right distance, which they dubbed Ceres and Pallas. But these objects were so small they could never be well imaged the way one could observe the surface of Mars or the clouds on Jupiter. Thus, they were dubbed "asteroids," again after the Greek (asteroeides, translated to "star-like"). We now know these asteroids are tiny hunks of rock and that there are over a million of them. All combined, they still wouldn't make an object big enough to be a planet, but their orbit is cursed by Jupiter to never be stable enough to successfully clump together.

What's especially nice about the discovery of asteroids is it demonstrates the ongoing quest of science to follow the evidence. The pattern in planet spacing pointed to something being between Mars and Jupiter. People searched for a planet and instead found tiny objects that fit some of the expected behaviors but not all of them. After initially calling Ceres and Pallas planets, their category was refined as more observations were made about them; they were downgraded from planet to asteroid. The understanding of asteroids as a small rocky body, essentially leftover rubble from the formation of the solar system, eventually took over from our understanding of them being "star-like." The process of science is always one of iteration—from discovery and misunderstanding to a slow but hopefully persistent gain of true knowledge. The original misunderstanding, the fact that not all the evidence fits the theory, is part of what drives discovery forward. It's amusing that we are always so quick to label things "star-like." Later in this book, you'll hear about objects that we still call quasars, a contraction of "quasi-stellar object." You may not be shocked to learn that quasars are nothing like stars (in fact, they are extremely bright galaxies).

Other examples of this iterative process litter the solar system. Our self-centeredness and, in some cases, regular lack of imagination, mean that this cycle of assuming a new thing is like something we already know nearly always follows the same course. Some planet or object is observed. People assume it's probably like the Earth—made of rocks, full of life and trees, and maybe has an intelligent civilization. More observations are made. Sometimes we send a robotic spacecraft, if we can, to look closely. Eventually, we realize that it is *nothing* like the Earth and feel semi-foolish for our original wildly inaccurate assumptions. Finally, we figure out that while *this* object is not like the Earth, we console ourselves by looking for the next thing that will definitely end up being like the Earth. "Always looking for another Earth" should be embroidered on pillows on the actual Earth. We are always searching.

Some fun examples of this are Venus and Mars.

Venus is the third brightest natural object in the night sky (when conditions are right, the actual third brightest object is the International Space Station, but that's unnatural). Early telescope observations of Venus explained this brightness: it's close to the Sun (two times closer than the Earth) and it's highly reflective. It reflects about 70 percent of the light that hits it, compared to only about 30 percent on average for Earth. Its incredible brightness means every culture on Earth made observations of Venus. The ancient Maya civilization believed Venus to be the most important celestial body and made careful observations of its motion, using it as part of the calendar system.

Venus through a telescope appears opaque, almost unknowable. The thick clouds that obscure Venus's surface are ever present and require precision on the part of the observer to note any details at all. Dusk and twilight observations are surprisingly the best time to observe Venus, when the cloud structure and faint, darker bands are easier to resolve. The first real observations of Venus that noted its clouds (all previous ones focused on observing the phases of Venus) started in the 1700s but didn't make any progress for a few hundred years. Astronomers were uncertain whether they were observing the planet or clouds. Some astronomers were convinced they saw dark patches, like the dark *mare* on the Moon, and calculated a day on Venus to be twenty-four Earth hours. Others were convinced there were long dark striations and that the day was 200 Earth days long.

Most photographic images of Venus showed a serene and placid sphere. Those taken in the early 1900s revealed a planet that stubbornly refused to give up any details about itself. Dark patches seemed to come and go mysteriously. In photographs taken in the near ultraviolet in the 1930s, the clouds showed interesting structure, but no one followed up on those early investigations. In the absence of details, imagination filled in the rest. If Venus was covered in thick clouds, it must have a tropical climate underneath them. It's closer to the Sun and so it's probably nice and warm. Maybe Venus was a tropical paradise full of life and activity. In science fiction from this time, Venus is a warm and lush destination.

This image wasn't dispelled until the 1960s with the first measurements of the temperature of the Venusian atmosphere. It was observed by both US and Soviet spacecraft and the atmospheric composition was found to not be full of water and oxygen, but instead primarily CO_2, and enough to create hot conditions at the surface. Additional observations, and eventually landers in the 1970s, revealed the truth—the surface of Venus was a hellish nightmare. Hot enough to melt lead (870 degrees Fahrenheit), a thick atmosphere creating a pressure equivalent to being 3,000 feet below the surface of the ocean, and clouds of sulfuric acid. The few spacecraft that landed on its surface had to work quickly; all failed within hours. We haven't sent a lander to Venus since 1985.

Venus is much more like Earth's evil twin than any other planet we know of. About the same size, with a thick atmosphere, Venus in the early days of the solar system may have been a tropical paradise, with liquid water and potentially oceans. But the luck of the draw in planet formation and a runaway greenhouse effect gave us the Venus we know today: dry, scorched, and barren.

More recent work has indicated that Venus may harbor life in its upper atmosphere, potentially single-celled organisms that float at temperature and pressure ranges that are more comfortable. Such signatures are not confirmed though, and despite us knowing more about Venus than we have ever known in all of human history, she still remains an enigma.

In any discussion of our solar system, one must pause at our misunderstanding of Mars. Another planet that we so hoped would contain life and civilization but ultimately has disappointed us. Early observations of Mars showed its polar ice caps (mostly CO_2 ice, a.k.a. dry ice, but a little water ice too) and Percival Lowell famously convinced the world that Mars had canals and other signatures of intelligent life. In the late 1800s and early 1900s, the obviousness of Martian civilization was so accepted as a fact of life that there were numerous science fiction books about it. Most famously, a 1930s radio play of the fictional story *War of the Worlds*, where Martians attack the Earth, caused a panic when people tuned in mid-play, thinking it was really happening. We've since (also) learned that Mars has never hosted a civilization and what Percival Lowell thought were canals could have been a slightly misaligned telescope or an eye condition. Mars is dusty and dry and thus far has not turned up a single Martian, if you don't count the many rovers we've sent to the red planet. It's also possible that like Venus, Mars, in the early days of the solar system, might have been a place with liquid water and maybe the building blocks of life. But due partly to its small size and distance from the Sun, Mars today has a thin atmosphere (mostly CO_2), creating wind and light dust storms but not an atmosphere thick enough to protect or support any lifeforms.

In our never-ending quest to find life, we are now convinced that some moons in the outer solar system may have liquid water—sub-surface oceans kept safe under an icy crust. The best candidate for this is a moon of Jupiter, first discovered by Galileo, called Europa. There are others around Saturn and maybe Uranus. It remains to be seen if those sub-surface oceans exist at all, and if they do, weather this is an alien octopus civilization living there. We can always hope.

Before we move on to other topics, we must of course briefly touch on my favorite planet, and honestly one of the weirder things we've observed in the solar system—Saturn and its glorious hexagon. Jupiter may have its red spot (a giant, centuries-old anti-cyclone), but Saturn has a north polar hexagonal vortex, first observed by Voyager in the '80s and seemingly still there. The hexagon is almost like a track, perfectly regular and rotating at a similar rate to the planet (we think—the exact rotation rate of Saturn is not perfectly known). Clouds run around the track, faster than the rotation rate of the hexagon. This structure may be the result of interactions between bands of winds at different speeds and latitudes in the Northern Saturnian hemisphere. The exact mechanism is unknown, but also weirdly, the color of the north pole of Saturn changed from blue to yellow in the 2010s, potentially the result of changing seasons on Saturn. The hexagon remains for me, the weirdest and most fun thing in our solar system.

Of course, many other fun and strange things are in our solar system. Uranus is tipped on its side, as if some jerk pushed it in the early solar system. The Moon is gigantic in size relative to the size of the Earth; no other solar system moon comes close in relative size to its host planet. If Jupiter were a bit larger, it could have been a star. Mercury is one of the reasons we knew there was a problem with Newton's conception of gravity—it wasn't until Einstein's general relativity that the orbit was correctly calculated.

As we will see in future chapters, our solar system is special not just because we are residing in it, but as we've learned more about other solar systems, we find that their structure, number and order of planets, and many other things are different from ours. We thought if we found other solar systems there would be variations on our system, but the truth is our solar system is pretty special.

What this means for us: The place in the universe that we know the most about is our cozy little solar system. One perfect star, eight planets, a grab bag of minor planets, asteroids, comets, and random space detritus, this solar system has everything a civilization could ever want. In fact, as far as we know, it's the only place with a civilization at all.

CHAPTER 9

Planets Are Surprisingly Common

Weird Facts: A lot of planets are around other stars. Weirdly for us, astronomers for a long time thought that planets were rare and unusual. But as we've gotten better instruments for finding them, it seems like nearly every star has a planet.

Our solar system has a ton of planets. We shouldn't be surprised that planets are common in the galaxy, and yet, here we are. As an elderly millennial, I remember when there were no known planets outside of our solar system. Then I remember when we had a few, their discovery in 1995[2] was a watershed moment in astronomy. Then I remember when there were ten, and 100, and now over 5,000. I will probably one day look back at even a count of 5,000 as a pathetically small number, demonstrating how little we knew in the past. Isn't it always that way though?

For nearly all of human history, we've known about five planets. The visible-eye planets of the solar system were named by the ancient Greeks—Mercury, Venus, Mars, Jupiter, and Saturn. These are so bright that they can be easily seen by the naked eye. With the invention of the telescope and photography, we identified the other large bodies in our solar system—the distant, cold ice giants Uranus (discovered in 1781) and Neptune (discovered in 1846), plus a slew of other smaller rocky and icy bodies—the largest of which are Pluto (discovered in 1930), Ceres (discovered in 1801), and many others discovered more recently.

But it would be 150 years from the discovery of the last new planet (Neptune) to the discovery of planets around other stars.

---

2   As you'll see in the next pages, this isn't technically correct, but people don't like the first planets ever discovered around another star and they skip over them.

First Discovery— Kind of a Letdown

The first planets to be discovered around another star were (are?) somewhat strange. You've already learned about the deaths of stars, which can be spectacular and explosive or calmer and quieter. In the explosive deaths of stars, remnants can be left behind—the condensed core of the star remains as a kind of cinder. This remnant, if it forms, is called a neutron star, the densest matter possible, basically a giant atomic nucleus. These neutron stars aren't truly stars—they put out some light but are not providing the warmth and energy of something like our Sun. Neutron stars can have intense magnetic fields, can make jets of energy out of their poles, and sometimes are rotating quickly. It's around this weird, dead, stellar remnant that the first extrasolar planets were discovered in 1992 (*not* 1995, as most people will claim, including me earlier in this chapter).

Three planets have been discovered around the neutron star (called PSR B1257+12), all of which are orbiting close to the star. One is the smallest planet we've yet discovered, at just 2 percent of the mass of the Earth. They were discovered because the orbital pull of these planets changed the timing of variations in the light from the star. The star is rotating quickly and predictably, and these little changes in the timing of when the star would get bright and dim perfectly matched three planets orbiting the star.

But this wasn't received by the scientific community as a huge discovery as you might think. These planets probably formed after the death of the central star, a second round of formation in a leftover disk of material after the explosion. In this sense, these planets were never exposed to the nurturing warmth of a sun, but instead have formed and existed in darkness, with a harsh, tiny, and faint star in their sky. Thus, the chances of life existing on these planets are minimal, and scientifically, we can't learn much about planet formation in general because they formed in such an unusual way. In the history of extrasolar planets, these three little planets are usually relegated to a footnote if they are mentioned at all. But they provided the first hints that planet formation may be common.

Second Discovery, or the One That Won a Nobel Prize

Despite having evidence that planets *can* form around the dead husks of old stars and thus maybe they should be abundant, there was still a feeling that finding planets around normal stars like the Sun would be too hard. In 1995, Swiss astronomers were the first to report the discovery of an unusual object—something about half the mass of Jupiter orbiting close to a nearby Sun-like star called 51 Pegasi. This possible planet had an orbital period of about 4.2 days—meaning a year on this possible planet was only 4.2 days long—not even a week. This discovery was published in *Nature*, titled: "A Jupiter Mass Companion to a solar-type star." The authors, Michel Mayor and Didier Queloz jointly won the Nobel Prize in Physics in 2019 for the discovery of 51 Pegasi b.

The planet was discovered using a method known as the radial velocity method. The motion of a star changes in subtle ways when there are other objects in orbit around the star. If the other object is big (like a companion star) then the effect on the motion is stronger. Similarly, if the other object is close to the star, the effect will also be stronger. One way to visualize this is to think of two people facing each other, holding hands, and spinning as fast as they can. Imagine one person is visible but the other person is invisible. You would see one person swinging in a way that they could never do if they were alone. The star swings back and forth, pulled slightly by the gravity of the planet. This measurement can be made by calculating the speed at which the star is moving either toward or away from us. The swing of the star causes the speed to wobble a little. These effects are subtle—generally the stars are massive and even a Jupiter-massed planet is dwarfed by the star. In the case of 51 Pegasi b (0.42 Jupiter masses), the star was swinging at about thirteen km/s every few days, right at the detection limit of the instrument that found it. In contrast, Jupiter causes a change in the velocity of the Sun of 12.4 m/s over twelve years, nearly 1,000 times smaller of an effect.

The discovery of 51 Pegasi b was the result of nearly a decade of searching for this signature. Several teams of astronomers had been scanning nearby, Sun-like stars, looking for planets. Why did the team of Mayor and Queloz find one? First, and maybe most importantly, they had a new instrument that could make the precise measurements needed. If they had used something even a little less sensitive, they would not have noticed the motion of the star; it would have been lost in the noise of their data. Secondly, they got super lucky. They found 51 Pegasi b during a survey of 142 stars. Other researchers were similarly looking at hundreds of other stars. It's possible that a survey of those stars today, when our technology can make measurements nearly 100 times more precise, that we would find planets there. But in 1995, it was like searching for a needle in a haystack. The third key to the discovery was that they found a weird planet, which turned out to be most easy to detect with the method they were using.

Why Is Every Jupiter Hot?

Well, not every Jupiter. But for the first few years of planet finding, astronomers discovered an incredible abundance of Jupiter mass planets orbiting against their host stars. In the sequence of discovery, the first, second, third, fourth, and fifth planets were the same. These planets were so close that they were puffed up from the intense heat, so close that they completed a full orbit around their star in a matter of short days, so close that they were evaporated away by their host star. These planets were all classified as "hot Jupiters" because of their masses and proximity to their stars. For the radial velocity method, this is the easiest type of planet to detect—the large mass and close distance mean the swing of the host star is large. But we weren't expecting to find them.

These hot Jupiters are strange because, in our solar system, Jupiter of course is not hot at all but majestically orbiting the Sun at five times the distance of Earth's orbit. Our theories of planetary formation up until these discoveries were all based on our solar system and didn't anticipate such a large planet so close to the star. Interestingly, there was one early paper doing computer simulations of planetary system formation. The authors kept getting strange results. Their model solar systems would form a Jupiter mass planet, and then the Jupiter planet would slowly migrate inward, the radius of its orbit shrinking—moving from the outer parts of the simulated solar system to right up against the host star. This migration is an effect of the interactions between the forming planet and the gaseous disk that the star and system form from. The authors were puzzled by this and noted in their paper that the simulation indicates that we should be asking ourselves why Jupiter didn't migrate inward and destroy the solar system.

If our solar system's Jupiter had somehow formed and then migrated to the location of all of these hot Jupiters around other stars, it would have destroyed the delicate orbits of all the inner solar system planets. No Mercury, Venus, Earth, or Mars—they would have been flung out into space billions of years ago. No life on Earth—which would be a cold, frozen world floating out in interstellar space. The fact that this didn't happen is why we are here—why you are reading this book. More recent theories speculate that Jupiter did migrate inward, getting to roughly the current orbit of Mars, and then turned around and settled down to its final location. This reversal was likely due to the influence of Saturn and the theory can explain some things about our solar system that are not well understood, like why Mars is so small relative to the Earth, and why Venus and Earth are similarly sized.

But we now know that not *all* planets around other stars are hot Jupiters. With the advent of other detection techniques that aren't biased toward massive planets close to their stars, we have found a wide range of planets, all at different distances from their host star with different masses and different orbital characteristics. This has yielded a massive archive of thousands of planets, discovered by telescopes both on the ground and orbiting in space, detected and verified by multiple techniques, and creating a rogue's gallery of strange and wonderful worlds. In the end, we think only about 1 percent of planets are hot Jupiters, an occurrence rate that puts our solar system firmly back in ordinary territory. And yet we still haven't found a perfect other Earth or Earth-like planet, floating out there around a star like our Sun.

Where Are All the Earth-Like Planets?

In answering this, we have to examine our observational limitations. While planets seem to be abundant around other stars, and while our technology has gotten roughly 100 times better than in 1995, the signature of the Earth orbiting the Sun is still barely detectable, about ten cm/s of swinging in the velocity of the host star over the course of the year. That's about 100 times smaller than the signature of Jupiter orbiting the Sun. To confirm a detection of a planet, that swing has to be observed for at least three years to rule out false positives. So observations of other stars' Earths require precise instruments 100,000 times better than the instrument that found 51 Pegasi b, and they have to carefully observe the star over three or more years.

Out of all the detected planets we know of, only six have masses less than five times that of the Earth *and* orbits longer than 100 days (all of them are super-Earths). One of the six orbits a dead star like the first planets ever discovered, orbiting a stellar remnant neutron star. Another is more like a mini-Neptune and likely doesn't have a solid surface. The remaining four are in varying stages of confirmation and need follow-up to understand their properties. Because these are some of the hardest planets to discover, it will likely be a few more years before we start to have meaningful statistics about how often an Earth-like planet in an Earth-like orbit around a Sun-like star occurs.

What this means for us: Most stars have planets, but most planets, as we will see in the next chapter, are nothing like the Earth and most solar systems are different from ours. Our current best guess, if you are being an optimist, is that potentially 1 to 2 percent of all Sun-like stars have an Earth-like planet in the right orbit that could have liquid water on the surface. But numbers being what they are, 1 to 2 percent is still a lot of planets. In our galaxy alone, this would mean around a billion or two Earth-like planets are floating out there. Hopefully we will discover at least a few of them soon and start to understand how common Earth-like planets and human-like civilizations might be. Why we haven't heard from any of them is a question for another chapter.

CHAPTER 10

Most Solar Systems Are Nothing like Our Own

Weird Facts: Almost none of the solar systems that we've discovered are like our solar system. We used to think every solar system was like ours (that's our human-centered view of things cropping up again). Instead, we've found some of the weirdest, most inhospitable planets orbiting everything from dim, tiny stars to stars bigger and brighter than the Sun.

The previous chapter may have given you an indication of this, but most of the solar systems that we have observed around other stars turn out to be nothing like our cozy and sweet little solar system. Instead, they have Jupiter mass planets all over the place, Neptunes instead of Saturns, and a conglomeration of extra-large, hulking super-Earths. Just a big mess. In a larger sense, it's a more interesting galaxy that way. In science, we almost always learn more when discoveries don't match our expectations and we are forced to revise our models and theories in light of a universe that doesn't comply with our preconceived notions.

Before we get to a few of the weird and wonderful planetary systems, let's first have a reminder of what our solar system is like. Our Sun (also technically known as sol) is an ordinary G-type star with a mass of one solar mass and a surface temperature of between 5,000 and 6,000 degrees Celsius. Stars like this are somewhat uncommon because the Sun is larger than average. Sun-like stars (G-type is a certain classification; other sizes of stars are classified with different letters) make up about 7 percent of the stars in the galaxy. Smaller stars, which are cooler and redder colored, are more abundant. Our Sun has eight planets—four small, rocky planets in the inner solar system and four larger gaseous planets in the other solar system, with a few different belts and clouds of rubble leftover from the formation of the solar system (the asteroid belt, where asteroids live and farther out, the Kuiper Belt and the Oort Cloud, where comets and things like Pluto live). The most massive planet (Jupiter) is 2.5 times the mass of every other planet added together and it's far from the Sun, taking almost twelve years to complete one orbit. Only one planet (Earth) is smack in the middle of the habitable, goldilocks zone of Sol—at just the right distance for a temperature that allows water to remain in a liquid state on the surface. A cozy solar system just right for us. In some sense, the Sol system is the only one that isn't weird.

But we know of thousands of new solar systems. Here we give a brief discussion of a few weird and wonderful ones, so different from our own.

Λ Hot Jupiter, Evaporating Λway

Our first weird solar system surrounds a star with a boring name. The star, HD 209458, is a bit more massive than our Sun. In many ways, it's like the Sun. HD 209458 is 157 light years away from us, in the constellation of Pegasus, and you can easily see it with a set of binoculars or a small backyard telescope. As far as we know, it has one lonely planet called HD 209458 b (all extrasolar planets are named after the star starting with b and then incrementing by one letter for each new planet). This planet is 0.6 times the mass of Jupiter, a gas giant made of mostly hydrogen, and it orbits its star every three and a half days. This means it's close to the star; at four million miles away from the star, its orbit is one-eighth the distance of Mercury's close orbit to our Sun. The orbit is so close, in fact, that the planet is being roasted by the star. Planets this close to their star are almost puffy—the heat from the star makes them expand like a marshmallow caught on fire. The expanded atmosphere, tenuous and superheated, is easily evaporated, giving rise to a tail of escaping hydrogen streaming behind the planet. This tail also contains other elements—carbon and oxygen—that get caught up in the evaporating hydrogen. This evaporation is strong enough that over the five billion years since HD 209458 b and its star formed, the planet has lost probably 7 percent of its total mass, a huge amount of material. The mass loss isn't enough to completely disintegrate the planet (although that can happen to planets and to moons of close-in planets), but the planet will continue to evaporate until the death of the star.

HD 209458 b is also interesting, not just because it's a puffed evaporating planet, but because it may have a magnetic field. Some planets in our solar system have magnetic fields (the Earth has an important one; Jupiter and the gas giants also have them, as well as tiny Mercury), created by their spinning, liquid metal cores. Magnetic fields provide a buffer between the activity of the star and the atmosphere of a planet and can help (in the case of Earth) protect life from particles from the star and, to varying extents, prevent atmospheric loss. We think there may be a magnetic field based on how hydrogen is evaporating from the planet.

Finally, this was the first planet found using a new method of finding planets. In 1999, astronomers used the "transit method" to detect this planet. This method uses the fact that if the orbit of the planet is oriented just right relative to the Earth, you can watch the planet eclipse the star. It won't totally block the star like a solar eclipse on Earth, but it can reduce the brightness of the star by a percent or two. HD 209458 b covers around 1.5 percent of the surface of the star during a transit and thus the light coming to us from the star drops by the same amount. If this dip repeats with a regular cadence, it might be a planet. In the case of HD 209458, it was. HD 209458 b was also the first planet to have its spectrum measured, telling us that the planet has incredibly fast super winds whipping across the planet, winds traveling up to 4,000 mph, stronger than any wind on Earth, Jupiter, or Saturn.

Many solar systems are like this one—a massive, Jupiter-sized planet nestled right against the star, evaporating away like a marshmallow on fire.

Λ Tatooine Solar System IRL

At the beginning of *Star Wars: A New Hope*, Luke Skywalker wistfully watches twin suns setting on his desert home planet, Tatooine. This incredible planetary arrangement is never remarked upon in the movie, Tatooine just being a place that Luke wants to escape from; in his own words it's farthest from the bright side of the galaxy.

In our own galaxy, we know of just twelve solar systems like this—where there are one or more planets orbiting a pair of stars. While most stars in the galaxy form in pairs or groups of multiple stars, the orbital mechanics of these systems make it hard for them to form planets. This is because the other stars in the multiple-star system typically disrupt the orbits of planets that formed close to just one star. Binary stars can only have stable planet orbits if the two stars are close together and the planets form around both stars at once. Or alternatively where one of the binary stars is so far out that the orbits never cross or interfere with the orbits of planets around just one star.

Out of the twelve confirmed circumbinary solar systems, where planets are around both stars, just two have multiple planets. One, TOI-1338, has two stars that orbit each other every fourteen days, a moderately tight binary system. The orbits of the two planets TOI-1338 b and TOI-1338 c are farther out, orbiting every ninety-five and 215 days, respectively. Both planets are gas giants, similar to Neptune and Saturn, and are probably too close to the stars to harbor life even on a moon orbiting one of the planets, so it's unlikely that any Luke Skywalkers are wistfully gazing at sunsets anywhere on TOI-1338 b or c.

But there are a few circumbinary planet systems that may have a contemplative Luke gazing at twin sunsets. One of them, and a nice example of this type, is Kepler-1647 b. While the planet is a gas giant a little larger than Jupiter, it is in the habitable zone of its two suns (both are about the size of our single Sun). If the planet had an Earth-sized moon, then the moon, rather than the planet, could have life on it. For many of the circumbinary planet systems that we've discovered, the planets are gas giants. It's possible that looking for habitable moons will be more fruitful in the search for life in these cases.

One interesting scientific angle for studies of circumbinary planetary systems is determining the order or sequence in which the planets and stars formed. In the case of nearly all circumbinary planet systems, it seems like the planets are orbiting in the same plane as the rotation of the two stars, which hints at a common origin—the planets and stars formed at the same time out of the same proto-planetary/proto-stellar disk. Misaligned planet orbits and star rotations don't seem to be as common as they are in single stars.

TRAPPIST—a Mini Solar System

The last planetary system we will examine is sometimes referred to as a "mini solar system," the TRAPPIST-1 system, with seven planets. The star, TRAPPIST-1, is a dim, small star, just 9 percent the mass of the Sun. Incidentally, this is almost the smallest mass star that can actually be a star, fusing hydrogen into helium. TRAPPIST-1 has seven planets all crammed close to the star— closer than the orbit of Mercury is to our Sun. The planets orbit the star in a matter of days—completing one orbit in anywhere from 1.5 days to nineteen days. Every planet is racing around the star. The planets are so close to each other that they likely create tidal effects as they pass each other in their orbits.

All of the seven planets are similar in mass to Earth, and likely rocky in composition (no Jupiters or even Neptunes to be found here). Three of them—e, f, and g—are within the narrow habitable zone of their star. But the conditions for planets so close to their star are harsh—they experience a lot more X-ray radiation from TRAPPIST-1 than we get from the Sun and are also within the magnetic field of TRAPPIST-1, which might make it hard for them to develop and hang on to an atmosphere. There are still possibilities for life around the TRAPPIST-1 system, but we need more observations of the planets to understand the conditions on each planet.

This system is a great case study, however, for how complex the requirements for life like our own actually are. While at first glance, a system with seven rocky planets should be more likely to have an Earth-like planet, we find that the star isn't suited for this and the orbits of the planets, while in the habitable zone, aren't necessarily helping anything. Such a cramped, small system ends up being potentially too hot, too crowded, and too hostile for even the development of a nice, warming atmosphere. While this discovery led to a lot of press coverage about a mini solar system, upon closer examination, it's nothing like ours.

Let's Not Over-Constrain Where Life Could Form

One thing to consider is how unusual our solar system seems to be. Several small, rocky bodies close to the star and several large, gaseous planets far from the star. A massive planet as the innermost gaseous planet (this is Jupiter), which acts as a shepherd and vacuum cleaner for all the leftover solar system rubble. An ordinary type star that isn't active.

We might be tempted to say that a solar system harboring life needs a Jupiter to keep all the riffraff of planetary system formation from hitting the life-harboring other Earth. We might also be tempted to say that other Earth needs a moon like we have our Moon. Our perfect little Moon does a lot for us—helping maintain the stability of Earth's obliquity (the tilt relative to the plane of the Sun, right now about twenty-three degrees) as it spins. The Moon gives us tides and may have helped foster the formation of life in tidal zones on the early Earth.

But quickly, you could end up in a situation where the only acceptable type of life-harboring solar system is one that looks like ours, with all the same circumstances needing to occur. There are only a few conditions that appear to be necessary for life.

What this means for us: Solar systems around other stars are weirder and more varied than we could have imagined. The seemingly high frequency of planet formation (most stars seem to have at least one planet) is balanced out by the fact that most planets and solar systems are nothing like ours. Despite this, I remain optimistic that while (as you'll see in the next chapter) a lot of planets are not friendly, we will eventually find many planets as comfortable (to their inhabitants) as our Earth is to ours.

CHAPTER 11

Every Planet We Reach Is Dead

Weird Facts: We've found over 5,000 planets around stars that aren't the Sun. Out of those, we haven't yet found a single Earth-like planet. Ironically, while we always assume everything in space will be just like what we've found on Earth or in our solar system, it seems like this is the one area where, so far, we are singular.

Humanity is extremely self-involved. As we have seen, at every opportunity, we assume that we are at the center (sometimes literally) of the universe. We thought that Earth was the only planet there was. We thought that our galaxy was the only galaxy there was. We even thought that our solar system was the only one. We thought both Venus and Mars had abundant life and advanced civilizations. We simultaneously believe so much in our uniqueness but also believe that the universe must be just like our home.

But the Earth is not like other parts of the universe or galaxy. It's unbearably special.[3] This remains true even if we can discover Earth-like planets around other stars because when one considers how difficult it is to get to other places in the galaxy, we can see that Earth is the place for us. This is something people only truly realize when they see it in the wider context. Astronauts in space describe a transformation that they undergo when looking down on the Earth. "The overview effect" is a transcendent feeling of awe and wonder and a feeling of connection to everyone on Earth and to the Earth itself. This doesn't happen for everyone but is especially heightened for the small number of astronauts who have been to the Moon—they are the only people in all of history who have seen the entire Earth floating in the blackness of space. I draw attention to this effect, because even if we know nothing about other solar systems, when we see our tiny planet against the emptiness of the universe, we know deep in our bones that it is incredibly rare and valuable.

---

3   Editorializing here to say that it's special, so far. There may come a time in the not-too-distant future when we discover an Earth-like planet around a Sun-like star. If we have good enough instrument to find one, that probably means there are many and it will be only a matter of time before we have a whole catalog of other-Earths.

"Earth Rise" from Apollo 8, photographed by Bill Anders

And in the few decades that we've been finding planets, it turns out that so far, the Earth is unique like no other planet.

As we've learned more about our galaxy, we've realized a few things. First, that planets are incredibly numerous around other stars. In less than three decades, science has evolved from thinking there are hardly any planets, and even if there were planets, they are too difficult to detect, having a running count of over 5,750 confirmed planets around other stars (as of the end of 2024). The exact statistics of what fraction of stars have planets is still unknown—there are so many variables and biases in our ability to detect planets that are difficult to tease out. But we think planets are common; they form all the time, and with 100 billion stars in our galaxy, there are at least tens or hundreds of billions of planets in the galaxy too.

The second thing we've learned is that a huge fraction of these planets is outrageously inhospitable to our style of life. Like Goldilocks, we find so many planets that just aren't right: Planets that are too massive—monstrous gas giants more akin to Jupiter and the outer planets; planets that are too hot—orbiting right next to their parent star, cooked to oblivion; planets that are frozen solid, exiled to distant orbits; planets that have atmospheres of toxic hazes.

As we learned in the previous chapter, our solar system is unusual (at least, given our current knowledge; time will tell if it stays that way). The position of the Earth relative to the Sun, the amount of light the Sun produces, the fact that the Sun is a single star, not part of a pair or group, the fact that Jupiter stayed in the outer solar system and didn't boot out all the inner planets a long time ago, all these things make our solar system somewhat strange.

Out of those 5,750 planets that we currently know about, a huge fraction of them would be hellish places to live. About 10 percent are around the size of the Earth (1.25 Earth radii or smaller). 105 of the 5,750 may be in the habitable zone of their star, but that's not necessarily the same 10 percent that are around the size of the Earth. Even in those thousands of planets we've discovered, there aren't any that are the mass and density of the Earth and located at just the right distance from the host star to support liquid water, and just of the right composition to possibly have water and maybe life.

Let's look at the history of Gliese 581c, a planet orbiting a small red star called Gliese 581. Gliese 581c was a media sensation when it was discovered. This was the first planet measured to be kind of close in mass to the Earth, and thus likely rocky, while also being in the habitable zone of its star, the region where the planet's temperature allows for liquid water on the surface. The star Gliese 581 is only about a third as large as our Sun and with a lower surface temperature, putting out around 1 percent of the energy that our Sun puts out. Gliese 581 has at least three planets orbiting it, with 581c being the third rock from its sun. So far, so good.

But upon further examination, 581c is different from the Earth. First, it's more massive—at least 5.5 times the mass of the Earth, although given measurement errors, it could be as massive as ten times that of the Earth. The mass on its own isn't a problem, but we have some indications that "super-Earth"-type planets like Gliese 581c may be more like mini-Neptunes than super-Earths. This would mean they have thick hydrogen atmospheres and are more like small gas giant planets. Until we have more information about the planet's composition and atmosphere, it's hard to say if it even has a surface like our planet.

It may also be *just* inside the habitable zone of its star. In this sense, it could be similar to Venus in our solar system, a little too hot, a little too near to the star. This marginal position might be obviated by various cloud or surface compositions that reflect the light from the star into space. An icy planet, for example, or one with a great deal of cloud cover, could maintain a reasonable temperature.

Finally, we turn to the orbit of Gliese 581c. It's closer to its star—orbiting a full "year" in just thirteen days. It also is tidally locked to the star, which means its day and year are the same.

Gliese 581c presents only one side to the star at all times (the same way we only see one face of the Moon), so one-half of the planet is roasted while the other half is in perpetual darkness. A challenging situation from the perspective of good weather.

I've spent nearly a full page talking about this planet that at first looks sort of promising as an Earth-like planet, but when examined more carefully, turns out to be a hellish, too hot and simultaneously too cold, chunky rock, maybe covered in a thick gaseous envelope. Certainly not a paradise like the Earth. And this happens over and over when examining the details of these smaller planets. We keep searching for planets like Earth, and instead we find duds.

Nearly every planet is likely dead. Sara Seager, an MIT professor and expert on extrasolar planets, describes most of the possible "Earth-like" planets we've observed as anything but. She classifies them as: "Rock Giants"; "Cannonballs"—basically a planet made of solid iron; "Gas Dwarfs"—tiny Neptunes; "Diamond Worlds"— pure, hot carbon; "Ice VII" worlds made of hot, pressurized ice; and "Carbon Monoxide" worlds. Planets where, if there is an atmosphere, it's incredibly thick, hot, and toxic. Planets where tiny diamonds fall like rain. Planets so close to their stars that they are volcanic nightmares. Planets whose orbits place them under such extreme positions that they are dissolving from the stress. Planets whose orbits are so eccentric and noncircular that they spend decades in the frozen distant extremes of their solar systems, ice balls that only feel the warmth of their sun every few eons.

A canonical "Cannonball" planet is Mercury, the smallest planet in our solar system. Mercury is mostly metal, with a density a little lower than that of the Earth (5.4 g/cm³ for Mercury, 5.5 for Earth). This density comes from a largely metallic composition, with probably a great deal of iron.

An exoplanet "Cannonball" is K2-229b, with 2.5 times the mass of the Earth at a slightly larger radius, orbiting its sun in fourteen hours. This planet is also probably similar to Mercury, but on a larger scale—with a mostly iron core and little mantle or rocky crust. K2-229b may have once had a thicker atmosphere but that could have been eroded away because it's so close to its host store. Or it could have been the result of a collision of two rocky, iron-rich planets. We still have a lot to learn about how these "Cannonballs" form.

In the list of planet types from Sara Seager, we can't pass "Diamond Worlds." These planets are typically not just perfect floating diamond in the galaxy but instead are rocky planets with a mostly carbon composition. Diamonds are made of carbon atoms in a perfect crystal structure, which can be made with pressure, heat, and time. Some of the planets we've found, like 55 Cancri e, have compositions indicating a lot of carbon and are expected to form diamonds more easily in the depths of the planet than Earth. 55 Cancri e is eight times the mass of the Earth and twice its radius, so the density and pressure of the planet are much greater. These diamond worlds are only diamonds if you cut through a surface of lava and/or graphite.

But if you're searching for diamonds, look other places—maybe in the atmospheres of Neptune-like planets. Diamonds, albeit tiny ones, are found commonly in asteroids, and we think they form easily in the atmosphere of Uranus and Neptune. These planets, unlike Jupiter and Saturn, which are mostly hydrogen, have abundant methane (which has carbon). Deep enough in their atmospheres, the temperatures and pressures are high enough to force carbon into the crystal arrangement characteristic of diamonds. Diamonds might be rare on Earth, but they are abundant in the universe.

One interesting gassy "cannonball" planet is fun to consider. TOI-4603 is a star larger than our Sun, and has a planet, TOI-4603b orbiting it every 7.25 days. This planet is a strange one, one of the densest we've ever encountered. While it's roughly the same size as Jupiter in radius, it's nearly thirteen times as massive as Jupiter. It's so weighty that it's nearly a brown dwarf (an almost-star that we learned about already) but is likely not massive enough to qualify. Its density is nearly three times that of the Earth, and denser than lead.

Basically, every planet we've found is either a gas planet with no surface, a hellscape with a lava surface, being roasted by the star, or otherwise inhospitable. Even when transported to these planets with a fully functioning spacesuit and plenty of oxygen, your odds of survival are minimal at best.

What this means for us: So, is all hope lost for us? Are we on a singular pale blue dot, a one-of-a-kind blue marble, perfectly unique in a vast galaxy? I refuse to believe we are the only one. As Carl Sagan said, "If it's just us, seems like an awful waste of space." The galaxy is so vast, and planets are so numerous, and we've only even known how to look for planets for a tiny fraction of the time we've been looking at the stars. Every year, our tools for detecting planets and measuring their properties get better and better. Every year, our understanding of atmospheric dynamics and stellar evolution gets better and better. Right now, I can easily explain the lack of detected truly Earth-like planets by the fact that if we were looking at our Sun and solar system from far away in space, we could barely detect Jupiter, never mind the Earth. If in fifty years, when our techniques and telescopes have improved, we still haven't found a real analog to our perfect Earth, then we should be worried.

But even if we do find a perfect match to Earth, it may be too far for us to ever hope of arriving and seeing up an Earth 2.0. We may only ever look up at it and wonder what the creatures on it are thinking. Are they out there, looking for the perfect match to their planets? Do they dream of finding more beings like them?

CHAPTER 12

Quasars, Rings, and Galaxies at the Beginning of Time

Weird Facts: The brightest things in the universe are ancient galaxies. These galaxies send energy across the universe like lighthouses, altering distant travelers to cosmic events thirteen billion years in the past. These galaxies are so bright that we used to think they were stars, and only after some years of exploration did we realize they are some of the most distant galaxies we can observe.

We've explored stars, planets, black holes, and cosmic distances. But let us now turn to the largest objects in the night sky—all the weird and wonderful types of galaxies. Galaxies are generally defined as large conglomerations of stars (millions or billions of stars) which are gravitationally bound together, along with dust and black holes and lots of free-floating hydrogen gas. Most galaxies fall into two categories: spiral or elliptical. Some are tiny, as small as 1,000 stars; some are massively large, having eaten many other galaxies. Some galaxies long ago stopped forming new stars and are full of old, red, small stars. Other galaxies are forming stars like crazy, creating new massive bright blue stars. Galaxies are everywhere in the universe.

One of the most startling pictures ever taken by humans demonstrates how abundant galaxies are. In 1995, after the optics on the Hubble Space Telescope were corrected, the director of the Space Telescope Science Institute decided to devote some of his observing time on the telescope to a study of distant galaxies. Hubble stared at a tiny patch of sky, just above the joint between the bowl and handle of the constellation of the Big Dipper, for ten days straight. The patch of the sky that was observed was partially selected because it was believed to be empty—the scientists who came up with the target strategy wanted to observe a relatively unoccupied part of space, a region without superclusters or known areas of large-scale structure. What they found was incredible. This tiny, boring part of the universe was home to over 3,000 galaxies. Some of the galaxies captured were the most distant objects known at the time and demonstrated that even empty parts of the universe are abundant with light and structure. This image came to be known as the Hubble Deep Field, and it was such a success that Hubble went on to do additional deep fields, including the Ultra-Deep Field (eleven days with a better camera than the first Deep Field) and the eXtreme-Deep Field (twenty-three days' exposure). The Hubble eXtreme Deep Field contains over 5,000 galaxies and is shown below. For context, this patch of the sky is tiny, about 1 percent of the total area of the full Moon. You would need thirty million of them to cover the entire sky. So, when you look at the image, consider that this view, of a relatively sparse part of the universe, is just one tiny part of the entire universe. Multiply this by thirty million to get a sense of how much the universe contains. It's nearly as difficult of a concept to hold in your head as the vast distances in space.

NASA, ESA, H. Teplitz and M. Rafelski (IPAC/Caltech), A. Koekemoer (STScI), R. Windhorst (Arizona State University), and Z. Levay (STScI)

Nearly every point of light in this image is a galaxy. There are only a handful of foreground Milky Way stars (which you can identify based on the diffraction spikes around them, like the one in the middle right of the image). Every point, every tiny dot of light, every collection of just a few illuminated pixels is an entire galaxy. Some are as big as our Milky Way (beautiful, majestic spirals) and some are even bigger (likely the red, ball-shaped ellipticals). But many are tiny, many are weirdly shaped, and many of them are little dots seen from a vast distance. The diversity and splendor of this image are awe-inspiring—the universe is so vast and so full of things. How many stars are in this image, how many planets, and how many possible forms of life might be contained in this tiny piece of the vast sky?

In the Hubble Deep Field, we find some interlopers from the earliest days of the universe. Some of these ancient galaxies are tiny and faint, but some of them are extremely bright. In fact, the brightest things in the universe are super bright galaxies formed in the early universe, just a few hundred million years after the Big Bang. We observe them now from across the universe and, because of the finite speed of light, it has taken nearly thirteen billion years for their light to reach us. During those thirteen billion years of time passing, the ancient galaxies have evolved and merged and are completely different than we currently see them. The light sent out of those galaxies thirteen billion years ago tells us a story of extreme conditions and somewhat unique circumstances to drive such a massive output of energy.

Let's closely examine one of these anomalously bright galaxies. They are usually referred to as quasars, which is a contraction of their true name, quasi-stellar objects. When first observed in the mid-1900s, these extremely bright objects appeared as points of light, like stars.[4] They also emit across a wide range of wavelengths, like stars, putting out light in X-rays, radio waves, infrared, and visible. But they clearly weren't stars—they were too faint, and the emission signatures didn't match any stars that we know of. But because they resembled stars but were not stars, they were dubbed sort-of-stars. We know now that they are active galaxies, and the points of light we see are massive amounts of emission coming from the heart of the galaxy—the supermassive black hole at its center. In most cases, the emission from the black hole dwarfs the actual brightness of the rest of the galaxy, and so these quasars appear point-like, with just a hint of fuzz from the rest of the galaxy.

In the early days of the universe, everything was closer together. Galaxies formed and quickly merged, two galaxies colliding and eventually becoming a single, more massive galaxy. These mergers happened more frequently because the overall likelihood of being close to another galaxy was high in contrast to the present day, where the expansion of the universe has pushed galaxies farther apart and mergers between two galaxies are less common. During the first few billion years of the universe, the rate of mergers was as much as ten times higher than it is now. During every merger, the stars and gas of the two merged galaxies will combine as stars slide past each other and eventually settle into stable orbits in the new galaxy. The two black holes from the original galaxies must also merge, spiraling into each other until finally meeting and coalescing. Out of this union, a quasar is born. The merged black hole alone is not responsible for the massive

---

4   There is an interesting similarity here to the discovery and naming of asteroids. They appeared as pinpoints of light and were called asteroids because they were star-like (aster meaning star). Essentially a quasar is a different object that is star-like but named differently. It's also amusing that an asteroid (a tiny rock) and a quasar (a massive, ancient galaxy with an active black hole) are named nearly the same but are opposite to each other.

amounts of light coming from quasars, although it does provide a lot of the necessary conditions.

Some amount of material from the merging galaxies will end up around the now larger central supermassive black hole. Stars and gas that back in their old galaxies were on perfectly fine orbits now find themselves on trajectories that end in the black hole. Most of this material won't go directly into the black hole but instead will create a massive disk structure around it, called an accretion disk. A star that was previously a sphere will become torn apart by the tidal effect of the black hole—the force on the near side of the star will be so much greater than the force on the far side that the star will get stretched and shredded (this is sometimes called spaghettification, different from the nuclear spaghetti of a previous chapter), ceasing to be an orderly object. Instead, the constituent atoms of the former star will go into orbit around the black hole (well back from that photon sphere we learned about earlier). These hydrogen atoms, along with any other stars or gas clouds that got too close to the black hole, will order themselves into a thin, dense disk of material. The black hole sips on the inner edge of the disk, consuming the material closest to it. Over time, the black hole will eat the entire disk, but its consumption rate is limited so the disk can last for a while. This type of disk was featured most prominently in the movie *Interstellar* with a black hole called Gargantua. Gargantua is surrounded by a hot red/orange structure, an accurate representation of the accretion disk.

You may be wondering why an infalling, shredded star creates a disk and doesn't immediately flow into the black hole. This is because one of the few fundamental rules in our universe is that angular momentum in interactions must be conserved. A star orbiting around a galaxy (as our unfortunate example star had been before the galaxy merger) has some energy of rotation. This energy is known as angular momentum and can be described as how much spin or rotation an object or system has in aggregate. This amount of rotation has to be maintained. This is a rule in our universe that we can observe and explain mathematically, but we don't have a deep understanding of why. But this requirement means that momentum in the orbital direction will be maintained. A previously large, disordered collection of atoms will become a beautifully ordered disk surrounding the black hole. As the inner edge of the disk loses angular momentum (the angular momentum gets transported outward to the outer edge of the disk, but the details of this are probably beyond the scope of this chapter), the material will finally fall into and be eaten by the black hole, eventually merging with the singularity.

OK, now after this detour into one fundamental rule of the universe (conservation of angular momentum) we finally come to the explanation for these brightest objects. When observing the light from a quasar, you don't see light from the merged galaxies. Nor are you observing light from the black hole itself (silly you, it's black). Instead, you see light from the disk of shredded stars surrounding the black hole but not yet consumed by it. The disk will become incredibly hot due to a combination of enormous friction between particles as they orbit the black hole and a conversion of energy from gravitational energy to thermal energy. As the particles make their slow way closer to the black hole, they fall deeper into the potential well of the black hole. This falling also releases energy, which powers the quasar. The accretion disk of a quasar is one of the most extreme environments in the entire universe, putting out more energy than our entire galaxy times 1,000. Just a single (massive) object around a single (massive) black hole can provide 1,000 times the power of a whole galaxy. These disks are relatively small—a few to tens of light days across (in contrast to the Milky Way, which is over 100,000 light years across). And within this small area, they provide the power of hundreds of billions of stars.

During the time that the quasar is active, the supermassive black hole can consume as much as a few Suns' worth of mass every day. A quasar can be active for up to a billion years, giving the supermassive black hole plenty of time to grow more massive. We think this accretion process is one key method by which supermassive black holes grow. Of course, this growth takes time, and there are recent hints that they must grow faster than just by accretion to reach the sizes we measure in the early universe.

A quasar is truly an astounding thing, and one we are lucky *not* to be seeing up close. Earlier I mentioned that our galaxy is boring and in a boring part of the universe. That's partly because there are no active quasars around us. The closest active quasar to us is about 800 million light years away, although we know of close to two million quasars spread throughout the visible universe. They are all far from us because, in the present day, galaxy mergers that can create the instabilities to make a massive accretion disk are rarer. In 4.5 billion years, when our galaxy merges with the Milky Way, the resulting merged galaxy may have a quasar for a few hundreds of millions of years. But if the center of our galaxy were a quasar, the energy output would be enough to sterilize whole star systems close to the black hole. The intense energy and radiation would likely be blocked by all the material between us and the center of the galaxy (about 26,000 light years away), so the Earth would probably be unaffected. The brightness of the quasar powered by our Milky Way's black hole would be somewhat limited as well, because, as far as black holes go, it's not spectacular. It's only about four million solar masses, while true quasar black holes are on the order of billions of solar masses. This doesn't take away from how awkward it would be for us to have a huge radiation and energy-emitting monster in the neighborhood, but at least we wouldn't all die.

Finally, as a fun fact, while most quasars are so far that they are hard to see without large telescopes, you can observe one from your backyard with just a normal, commercially available telescope. This is 3C 273, in the constellation of Virgo. 3C 273 was the first quasar ever observed, and it's the brightest one as seen from Earth. It's just 2.5 billion light years away, and the black hole powering the quasar is nearly 900 million solar masses. You can find it with a six-inch telescope, where it will look like a star (after all, it's a quasi-stellar object), a massive burning disk of fire, orbiting a relativistic behemoth growing larger every day. A fun thing to look at with your eye and likely the most distant object that can be observed with a backyard telescope.

Quasars may be the weirdest thing a galaxy can host (certainly the brightest), but luckily for us, there are a few other weird and wonderful galaxies. The most fabulous is the fabled Ring galaxy. This is a class of galaxies that, unsurprisingly, look like rings. Let's examine.

NASA, R. Lucas (STScI/AURA)

It's a ring. Isn't the universe amazing? This galaxy is called Hoag Object and is the canonical ring galaxy. When these types of galaxies were first observed, there was a reasonable amount of confusion about them. Maybe there were two galaxies in this image: the central part is a different, more distant galaxy and there happened to be a nice apparent alignment. In fact, both the ring and the central core are parts of the same galaxy. To make things even more extra, a second ring galaxy is in that image, just above and to the right of the core, at around one o'clock. It rings all the way to the edge of space.

How do these galaxies form? While the perfect symmetry of this ring galaxy makes it look highly unlikely to happen in nature, there are a few ways to create this stunning galaxy. One way of making these rings is as a natural byproduct (like quasars) of galaxies merging. But it takes a particular galaxy merger to get this perfect ring effect. This type of merger is called a "bulls-eye" collision, where a smaller galaxy hits a larger spiral galaxy dead center, plunging through the galaxy like a rock thrown into a pond. The ripples from this will push outward, sweeping up any material in the way. The effect of the collision pushes the spiral arms of the galaxy outward, where they pile into a ring (this process takes time, so we likely see the above galaxy a few billion years after the merger, when the ring is well-formed). Pushing so much material (including diffuse clouds of hydrogen and other non-star stuff) into a smaller volume also sets off new star formation, as gas and dust get condensed into star-forming clouds. This is partly why the ring appears blue in color—all the newly formed stars give off bluer light than old stars. The central core is yellow, indicating no new stars have formed in the last few hundred million years, and likely is made up of older, smaller stars. These lower-mass, redder stars survive longer than big, massive blue stars. The colors of different parts of the galaxy tell us about their histories.

Other ring galaxies may have formed on their own—from gravitational instabilities in their own structures that cause the central part of the galaxy to physically detach from the outer spiral arms. These "self-made" ring galaxies are usually not as striking as those that result from mergers.

Many ring galaxies are not as well-ordered as Hoag's Object—some of them are still in the middle of a merger, and the ring and core have not yet separated. Some of them are slightly off-axis, without perfect symmetry. Some of them are caught in the act—

with the smaller galaxy having just passed through the larger galaxy. There are a small number of ring galaxies where the ring is misaligned to the rest of the galaxy—"Polar Ring Galaxies." Imagine a merger happening not as a bulls-eye but as a perfect hit directly into the side of the galaxy. Such a merger may result in a ring that is looped around the galaxy perpendicular to the direction of the galaxy's disk. These are even rarer than normal ring galaxies and we still don't understand the mechanics of their formation.

We know ring galaxies have been forming for a long time. In a vast universe full of galaxies, it's inevitable that some of those mergers will be perfectly aligned to create such beautiful outcomes. One of the most distant ring galaxies we've observed is nearly ten billion light years away from us, its constituent galaxies having been formed shortly after the Big Bang. Ring galaxies are a nice reminder of how, in a universe large enough, nearly all possible structures will be formed.

Quasars are a beacon to us from the distant and early universe. A legacy of the past when galaxies were brighter, everything was closer together, and the level of activity in the universe was higher. It is also interesting to explore briefly what galaxies were like in those early times, in contrast to their behavior and structure in the present day.

The earliest, most distant galaxy we've observed so far was recently discovered. It's called JADES-GS-z14-0, observed as part of the JWST Advanced Deep Extragalactic Survey (JADES survey, partly inspired by the Hubble Deep Fields). This survey is trying to find the most distant galaxies to understand the formation of the first galaxies when the universe was a baby. JADES-GS-z14-0 was discovered in 2024 and is the most distant and youngest of many galaxies we will observe from the baby universe.

JADES-GS-z14-0 is a blob of a galaxy, visible to us as a small clump of pixels as it would have appeared when the universe was 300 million years old. A few galaxies have been observed similar to JADES-GS-z14-0, less distant, but they all share similar characteristics. They are small: a few hundred to a few thousand light years across[5] with total masses around ten million times the mass of the Sun. They are creating stars at a huge rate: fifty to 100 solar masses worth of stars per year, almost 1,000 times more than the Milky Way's rate of star formation, despite their size. The overall picture of these early galaxies is that they are compact, early in their lives, and making stars at a frantic rate. Unlike a quasar, most of the light we detect from these early galaxies is from the stars themselves rather than the central black hole.

What's most interesting about these early galaxies is that according to our models of galaxy formation, they shouldn't exist, yet we observe them. So, our models must be wrong. We assumed that it would take nearly a billion years for enough material to build up in a single place for a true galaxy to form, as dark matter and primordial hydrogen slowly condensed under the influence of gravity. It seems like that process may take a few hundred million years, maybe less, with the first galaxies now visible and bright after just 300 million years. The quick

---

5 The Milky Way is about 100,000 light years across, so these galaxies are a few percent of the Milky Way in size or even smaller.

condensation of material is surprising, and like most things in astronomy, needs to be studied more. JWST, the largest telescope NASA has ever put in space, has been a key platform to detect and understand the early universe. As we observe even more galaxies from the early universe, we can finally start to figure out the rules for making them.

What this means for us: While it may seem like galaxies that are thirteen billion years old don't have too much to do with life here on Earth, they are a critical part of understanding a fundamental human question: *How did we get here?* Figuring out how the first galaxies formed is the first step to answering that question. Make the first galaxies, and eventually you make the Milky Way galaxy. Make the first stars, and eventually you make our perfect star, the Sun. While many of these questions, especially in relation to such distant objects as quasars in the early universe, can feel totally esoteric, they matter to you and me. Our fundamental understanding of both how we got here, and the nature of the universe could matter in a way we can't even begin to understand. The fundamental laws of the atom turned out to matter to humanity, giving us nuclear power and the atom bomb. We haven't yet touched the fundamental laws of the universe, but it may matter to us once we finally know them.

CHAPTER 13

Speaking of the Beginning of Time...

Weird Facts: The Big Bang was neither big nor a bang. It was tiny and silent, but the Tiny Silent is not a great way to start off a universe. At one time in the distant past, the universe was condensed into a single, infinitesimally small point of infinite energy and density. That point exploded outward (why? We don't know) and 13.7ish billion years later, here we are.

Unlike how the actual formation of the universe went, only now, two-thirds of the way into this book, do we get to the start of things. The Big Bang and the beginning of time.

The beginning of our universe is shrouded in mystery. The mystery is both literal and metaphorical. We can't see that far back in time (more on that later). But more importantly, the conditions that we believe existed in the first second of the universe are so weird and challenging that our physical models break down and stop working. This may sound familiar—a weird, extreme environment that isn't directly observable. That description also matches a black hole, and in some sense, there are many similarities between the physics of black holes and the Big Bang. In both cases our ability to model the situation breaks down because of the vast amount of matter and energy being compressed into essentially a space of zero volume (division by zero is tough for math no matter where it happens), but the breakdown in physics at the beginning of the universe happens for different reasons than with a black hole. The Big Bang Theory (physical concept, not the TV show) was developed over decades to explain multiple, different observations of the behavior of distant galaxies, other observations of the universe, and to try to understand why our universe is so weirdly uniform.

The original term "the Big Bang" was coined by Fred Hoyle, a British astronomer, in 1949 on a BBC radio program, saying: "These theories were based on the hypothesis that all the matter in the universe was created in one big bang at a particular time in the remote past." The term didn't come into general use in the world of astronomy until the 1970s. Ironically the term is wrong on both counts—the start of the universe was technically the smallest thing possible—the universe had zero size—and occurred in a medium without the ability to transmit sound and thus should be called the "Tiny Silent," but somehow, I don't think that will catch on.

So, let's begin at the beginning. One of the most important innovations of the Big Bang Theory is that time has a starting point. In earlier conceptions of the universe, it was eternal and static— the universe had always been and always would be, neither growing nor shrinking. Alternative views were that the universe had a start, for example, as described in the Bible, and was only a few thousand years old. This view was directly contradicted by the discovery of the fossil record on Earth and eventually by our understanding of geological time.

The eternal and unchanging "steady state universe" has its roots in early static models of the universe, where it was generally assumed to have existed forever. One question posed as early as the 1600s by Johannes Kepler (who figured out how orbits work) was, if the universe is infinite and has existed forever, why is the night sky dark? This question is sometimes referred to as "Olbers' Paradox," named for German astronomer Heinrich Olbers. If the universe is filled with stars and galaxies in all directions, given enough time, then the light from those stars should occupy every point in the sky, lighting up the night sky as much as the day sky. Since that's clearly not the case, both Kepler and Olbers came up with reasons for why the night sky is dark. Kepler theorized that after a certain distance, there are no more stars, while Olbers thought that space gradually absorbs light from stars and so at a certain distance out, all the light is absorbed. Interestingly, poet and writer Edgar Allan Poe suggested maybe the answer is the universe isn't old enough yet. In one of his last works, the prose poem "Eureka," he writes about Olbers' Paradox: "supposing the distance of the invisible background so immense that no ray from it has *yet* been able to reach us at all." Poe's "yet" contains an idea that not enough time has passed for stars to outshine the dark. However, it's quite a leap from these early fumbling guesses about the nature of the universe to our current understanding.

To get from Poe's *yet* to our current understanding, we have to also explore our own place in the universe. As described in an earlier chapter, we always assume that what we see and experience is all that there is. In our early understanding of the universe, there were many arguments about whether our galaxy was the only galaxy and thus, contained the entire universe. Only when we were able to measure distances to other galaxies did it become clear how massive the universe was. That what we called "spiral nebulae" were individual galaxies, each as massive and full of stars as ours. In addition to measuring the distance to Andromeda, Edwin Hubble noticed in 1924 that all of the galaxies he observed were not just distant but seemed to be moving away from us, with the more distant galaxies moving away from us faster. This link between distance and recessional velocity is now referred to as Hubble's Law and first ignited the possibility that the universe was not static but was expanding.

Because no discovery is ever the work of just one person, we would be remiss here to not describe the work of Alexander Friedmann, who predicted an expanding universe a few years before Hubble found it. Friedmann's work developed equations based on general relativity with a few important assumptions. The first assumption is that the universe is, on large scales, perfectly uniform—the distribution of matter is the same—and that this is true in all directions—there is no "center" or preferred direction or point. All points are the same and homogeneous. This concept is known as the cosmological principle and says no single location is special, and all parts of the universe are functionally the same. Making these assumptions, Friedmann determined that the behavior of spacetime depends on the overall mass density of the universe, creating a dynamic and changing universe instead of a static one. His equations allow for a range of universe behaviors, depending on what is inside the universe. A theoretical universe with a high mass density may crunch in on itself, while one with low density may expand forever. These equations are now used to determine the fate of our universe, but at the time were an invigorating application of general relativity on a universe scale.

Physicist Georges Lemaître, immediately upon learning that all galaxies are moving away from all other galaxies, suggested in 1927 that the next obvious step is to consider what happens if you reverse time. These now-distant galaxies must have once all been much closer together. Taken to an extreme, go far enough back and they must have occupied the same space. The universe in the past must have had a higher density and temperature than the universe of the present day. Lemaître proposed in 1931 that the entire universe started as a "primeval atom," containing all of the universe's mass and energy. This was the first real conceptualization of what we now know as the Big Bang.

At the time, multiple other theories of the universe floated around in a rich environment of discovery in physics, especially in the world of quantum mechanics. The scientists who didn't like the universe having a primeval atom starting point had a whole menu of universes to pick from. Oscillating universes, universes where light gets tired, empty universes, pick your favorite universe. But none of these rose to the importance of the dichotomy between a steady state universe versus the Big Bang universe argument of the mid-twentieth century.

The beginning of the end for the steady state was radio astronomy, which advanced significantly during World War II. While some preliminary measurements of the density of radio sources in the universe hinted that the Big Bang was right, the nail in the coffin was the discovery of the cosmic microwave background. The cosmic microwave background was discovered by accident, as a persistent background noise in radio measurements being conducted by Arno Penzias and Robert Wilson at Bell Labs in New Jersey. They were attempting to make incredibly precise measurements of radio signals from satellites, as well as astrophysical targets, and had worked hard to eliminate all sources of noise in their receiver. Despite extensive attempts to reduce the noise (including cleaning pigeon nests and droppings inside the horn of the antennae), they could not eliminate a mysterious and stubborn source that appeared to come from all directions, appeared both during the day and at night, and which was incredibly even. They concluded that it was coming from outside of our galaxy, but it wasn't until a friend of Penzias told him about a new paper from Princeton predicting leftover radiation from the start of the universe that the two realized the source of the noise. They later went on to win a Nobel Prize for discovering this trace signature of the Big Bang.

They had measured the smooth, perfectly even, perfectly isotropic remnant of the Big Bang. The universe was once hot and as it expands, it cools off. Any object with heat will give off photons (you can see the photons generated by the heat of your body with an infrared camera or when a poker glows red-hot after being in a fire). The current universe has cooled enough in the present day that the temperature is so low it gives off light with a wavelength on the order of millimeters. For comparison, visible wavelength light has a wavelength of hundreds of nanometers; infrared light has wavelengths of micrometers. The wavelength of the photons can be linked with a temperature, and so we can measure the background to be exactly 2.72548 degrees above absolute zero. The temperature of the universe has been measured so precisely that all of those digits I included are part of the measurement; we know it that well. Deviations from this temperature are on the order of one part in 25,000—this background is the smoothest, most even surface ever measured in astronomy. The discovery of this background was the end of the steady state theory of the universe and the real triumph of the Big Bang. With the theory of a steady state universe at that time, there was no way to explain an evenly smooth background, no way to explain why regions of the sky on opposite sides of our visible universe should have the same background, and no reason there should be a background at all. There were some attempts to add an ad hoc explanation for the CMB to steady state theories, but the power of the Big Bang Theory was that it predicted such a background before it was discovered. Adding it into a theory after the fact is less convincing.

But scientists are never satisfied with a theory—it must always be examined and explored further, poking at the edges to see where the theory stops working. Thus, for the Big Bang, there are other ways to test it and still more mysteries to be worked out. One test of the theory is based on the fact that for about twenty minutes at the start of the universe, enough time to make breakfast and a cup of coffee, the universe was the right temperature to fuse hydrogen into helium. Essentially, the entire newborn universe was the temperature and density of the core of a star. During this time, the universe created a handful of light elements—deuterium (hydrogen plus a bonus neutron), helium, helium-3 (helium minus a neutron), and the teeniest amount of lithium. This process is called nucleosynthesis. The ratio of hydrogen to helium post-Big Bang nucleosynthesis is about 3 to 1 by mass, so the normal matter content of the primordial universe is 75 percent hydrogen and 25 percent helium by mass, everything else being trace amounts but still detectable. By particle count, it's more like 92 percent hydrogen and 8 percent helium (twelve hydrogen atoms for every one helium atom). Astronomers have since measured the actual abundances of these light elements out in the universe, finding good agreement with the theory. Another check mark for the Big Bang.

Other tests (baryon acoustic oscillations, the distribution of galaxies and large-scale structure through the universe, measurements of the composition of primordial gas, and others) are too much to explain in these short pages, but all of these independent tests combine to tell a compelling story of the beginning of time.

In the last two decades, a twist in the story of the universe has been uncovered. When we look back into time, we see galaxies all moving away from us—this much has been known for 100 years. But only with more accurate measurements of the distances to those far off galaxies have we discovered that not only are they moving away from us, but the speed at which they are moving is growing. The more distant a galaxy is, the faster it's moving away from us, with exponential growth—the most distant galaxies are much faster than if you scaled the speed based on the nearby galaxies. This cosmic acceleration was discovered using observations from the Hubble Space Telescope and many ground-based telescopes and we only had the precision in measurements to notice it in the last twenty to thirty years. Not only is the universe expanding from the initial burst of energy from the Big Bang, but the expansion is growing, driven by an unknown force that may be part of the fabric of spacetime itself. The nature of our universe is much weirder than we previously thought (more on this accelerating expansion in a later chapter).

Some other elements of the Big Bang are not yet understood. These problems—the flatness problem, the horizon problem, the monopole problem—can all be summarized as problems of uniformity. Why is our universe so uniform, even, and smooth? Why should the cosmic microwave background be *perfectly* even? Why should the density and structure of one side of the universe be the same as the other? One solution to these problems is to have the early universe undergo a period of extreme, rapid expansion known as inflation, where the size of the universe expanded exponentially, essentially enlarging a tiny segment of the early universe into the entirety of the universe we see today. Thus, any small deviations get "inflated away," and you end up with a beautiful smooth universe where everything is the same—all features are derived from the properties of that one tiny patch of universe. A period of inflation has some of the same hallmarks as cosmic acceleration, and in both cases, the cause is not understood. For inflation, the whole process must happen nearly simultaneously with the start of the universe and must impart a great deal of energy into the early universe, as the inflationary field decays. If this seems contrived, that's because it is. Our understanding of inflation is limited by the fact that it takes place well before our physical models can start explaining the universe.

So, what was the early universe like? If the universe has a starting time, essentially a t=0, this leads to our first problem. Zeros are difficult in math—a division by zero creates infinity and infinities are hard to deal with. The Big Bang started in a singularity because of this zero; temperatures, lengths, and densities must go to infinity, which is unphysical. There may be something before the singularity, but as of right now, we can't know what it might have been. The singularity itself is also a problem for us. Our current quantum mechanical model of the universe has certain limits—the smallest size one can measure is called the Planck length, about twenty orders of magnitude smaller than a proton, or 1.616×10^{-35} m. Lengths smaller than this can't be understood well because of quantum uncertainty. Similarly, there is a minimum time interval that can be measured, the Planck time, 10^{-43} seconds. Lengths of time shorter than this can't meaningfully be measured, again because of uncertainty. There is also derived a Planck temperature, about 10^{32} Kelvin. This is a sort of "absolute hot" the way that 0 Kelvin is absolute zero, the lowest temperature something can be. Temperatures above this temperature can't be measured or meaningfully described because our physics doesn't allow it.

The early universe seems to violate all these Planck limits— temperatures above the Planck temperature, times smaller than the Planck time, and sizes smaller than the Planck length. This is the real problem with describing the early universe—our tools of measurement, the fundamental concepts of doing science like time, length, mass, and temperature, don't work in this extreme environment. The earliest seconds of the universe are shrouded in mystery because we can't model them. Inflation may or may not happen in these early moments. A fun fact is that during this time, the entire universe was dense enough to become a black hole (such an event would have strangled the early universe in its cradle), but the universe was expanding so rapidly that it escaped this fate.

After inflation, the universe continues expanding, although not as rapidly, and cooling off because of the expansion. It passes through phases defined by temperature—when quarks (subatomic particles) can first form, when the abundance of matter beat out the slightly less abundant anti-matter, when subatomic particles first form, when the fundamental forces decouple from each other, becoming individual forces (although gravity has never cooperated with this part of it). Finally, the universe is cool enough for hydrogen to fuse into helium (this is about thirty minutes in, when the universe is the temperature of the core of a star). Then it cools even more, taking a few hundred thousand years for atoms to form. The cosmic microwave background is an imprint of the early universe at around this time, about 380,000 years after the Big Bang, when it cooled enough to become transparent. Before this, the universe was hot enough that it was foggy with energy, and so the cosmic microwave background is a snapshot of what is sometimes referred to as the "surface of last scattering," after which photons can move freely through the universe without getting absorbed by hot plasma. This surface is the furthest back we can directly observe—everything earlier than 380,000 years after the Big Bang is inferred through its effects or by theory alone.

Eventually, under the ever-present influence of gravity, the newly formed atoms condense and cool enough for the first stars to form. The first galaxies form, and the universe takes on a shape that we can see and measure. Our galaxy forms, our star forms, and eventually 13.7 billion years after the start, we evolve to ask questions about it.

What this means for us: Our understanding of the beginning of time and everything is still missing key parts. We've filled in some of the picture, but similar to how our understanding of black holes is incomplete without better understanding gravity on quantum scales, our theory of the beginning of the universe is waiting for physics to catch up. But in these gaps, there is an incredible opportunity. Figuring out the conditions of the early universe might tell us more about the nature of spacetime and about the energy source that drove the universe's expansion. There is always an incredible opportunity for progress in fundamental physics, and the conditions of our weird universe provide the most extreme laboratory there is.

CHAPTER 14

The (Tragic and Yet Possibly Quite Boring) Heat Death of the Universe

Weird Facts: The universe right now is as bright as it's ever been and getting dimmer. We are living in the age of stars—trillions of years in the future, the last stars will form. But even now, the amount of star formation happening in the universe is less than it was even a few billion years ago. The universe is relatively young, and may already be past its prime.

We learned that the universe is expanding, and that the expansion is accelerating. This means nearly all galaxies in the universe are moving away from each other. This isn't because the galaxies have a lot of velocity or motion, but because the space between them is getting bigger. Any galaxy is pulled along by the flow of spacetime, although from the perspective of someone inside any one galaxy, it just appears that all other galaxies are receding away from their one. Space is getting bigger and all galaxies are riding that flow but without any additional mass or material being created. This flow, its speed, and how it may be changing with the history of the universe, is one of the primary clues for us about the nature of spacetime and the eventual fate of the universe. Taken at face value, if all galaxies are flying away from each other, then, depending on how quickly they are doing so, there will come a time when all galaxies are infinitely far apart and unobservable. Depending on the nature of the flow and how it changes, such an expansion could eventually tear galaxies apart, isolating individual stars. Eventually everything is so far apart that nothing can form, and nothing ever happens again. Is this sad fate to be the end of our universe?

When considering possible fates of the universe, the total amount of energy contained in the universe sets its eventual end. The Friedmann equations, which predict the overall behavior of the universe on its largest scales (and which we talked about in the previous chapter in the context of the expansion of the universe), have several different possibilities, all dependent on the overall density of the universe. Essentially the question to be answered is: is there enough mass in the universe to exert enough gravitational force to overcome the inertia of expansion set off by the Big Bang?

If the universe contains a great deal of mass, much more than our current universe seems to contain, the expansion that we observe could slow down and stop. If the overall density of the universe was high enough, the expansion could reverse, with the force of gravity overpowering the expansion and driving all galaxies and matter to move toward each other, inward instead of outward. Contraction instead of expansion. This scenario is referred to as "the Big Crunch" and indicates the universe is closed. The crunch ends with everything being destroyed into a single point of infinite density—something akin to the singularity that we think started the universe. A variant of the Big Crunch universe is one in which, after crunching, the universe bounces back—a Big Bounce. Refreshed, rejuvenated, and ready to start a new cycle of birth and death. Such a cyclic universe could cycle between expansion and contraction, bang and crunch, forever into infinity. The oscillations between a Big Crunch and a Big Bounce could go on forever, a cycle of rejuvenation and reincarnation. But alas, this does not appear to be our universe.

Another possibility revealed in the Friedmann equations is a coasting universe. This is a universe in which the expansion continues forever, but at an unchanging rate, just cruising throughout the eternity of time, getting bigger in a predictable way. A variant of the coasting universe is one where the density is just right and the universe never contracts but is slowing down such that at some infinite time in the future, it will stop expanding. This universe is balanced on a knife edge; a bit more mass makes it crunch, while a bit less mass makes it expand forever. Unfortunately, we exist in neither of these universes—no coasting for us.

Interestingly, these universe variants (the Big Crunch and the coasting universe) would all be much younger than our current universe. When we study the type of universe we live in, we can rule them out by figuring out the energy density in our universe, but we can also rule them out based on our measurements of the age of the universe. At our current measured age, there is no way we live in one of these universes; there just isn't enough stuff and too many galaxies are older than the ages that these types of universes would need. Instead, we live in what appears to be an expanding universe with an ever-increasing rate of expansion.

The accelerating expanding universe is one without an end and continuously getting bigger. The rate at which it is getting bigger also appears to be growing (this is what distinguishes it from the coasting universes). In the future, galaxies will be moving away from each other even faster than they are now, and eventually galaxies that could once observe each other become far enough apart and moving at fast enough speeds that they will no longer be observable. Space will stretch the wavelengths of even high-energy photons into undetectability. Galaxies will pass out of range. One way to imagine this is to consider galaxies just within our horizon, the volume of space that we can observe given the speed of light and the age of the universe. With a strong enough expansion, a galaxy that was once just inside the horizon could be pulled outside the horizon, receding from our view. Depending on the nature and history of the expansion, it could even be pulled outside any future possible horizons no matter how long you wait, never to be seen or heard from again. This cosmological horizon is the ultimate arbiter of what is or will ever be observable. We already know the area outside of our current horizon is likely an infinite universe that we will never explore or even see. The future accelerating expansion will move things that are currently inside our horizon to push them outside. A future alien civilization on our own galaxy, observing the universe at some much later date from now, might never know about those galaxies—to them, they don't exist and would never be observable.

This cosmic horizon is a much more sinister vacuum cleaner than any black hole could ever be. Whole galaxies, whole histories of the universe, will be lost across it. Pulled inexorably as much as if they were being destroyed in a horror movie. There is no force or chance or change that can outsmart this ultimate grim reaper. Some galaxies that we have observed—distant ones—have already been pulled across it, never to be seen again.

Taken to an extreme, this type of universe ends in what is called "the Big Rip," where the expansion becomes so strong that, after separating all galaxies from each other, it successively overcomes gravity at smaller and smaller scales. So next it pulls apart galaxies, then solar systems, then stars and planets, then black holes, and then even particles. Everything is ripped apart, even spacetime itself. We don't yet know if we live in a Big Rip universe. The exact nature of the accelerating expansion is still a mystery.

What we do know about our universe is that it's entering a period of accelerating expansion, similar in some ways to the strange inflation of the early universe. What is driving this acceleration is unknown, and we refer to it colloquially as dark energy (more, well, as much as we actually know, on that in the next chapter). We measured the overall expansion of the universe by measuring how quickly galaxies were moving away from us compared to how distant they were. Hubble did this in the 1920s, finding all the observations were best fit with a line. More distant galaxies were moving away from us faster than nearby galaxies. The slope of this line is called the Hubble constant.

But measuring distances in astronomy is tricky—stars and galaxies can have a huge range of masses and thus a huge range of brightnesses. Imagine a galaxy out in space, of apparent brightness when observed from Earth. Are you looking at a faint but super close galaxy or a bright but distant galaxy? One of the central problems in cosmology is understanding precisely how distant a galaxy is, typically by figuring out how bright it should be and backing out how far away it is. Alternatively, if you can find a particular object that is always the same brightness, you can use this "standard candle" to determine its distance. Astronomers have found what seems like a good (but not perfect) standard candle—exploding white dwarfs whose mass just exceeds the Chandrasekhar limit. White dwarfs are held up by the degeneracy pressure of electrons (as we discussed in a previous chapter), but this can only support a white dwarf up to 1.4 times the mass of the Sun. A white dwarf that tiptoes over that limit becomes unstable and explodes in a supernova. Unable to support its weight, it collapses and bounces back in a spectacular fireball. Every exploding white dwarf is the same—the same mass, the same material, the same brightness. Explosions from this type of white dwarf are called Type Ia supernovae. Type Ia's have a characteristic spectrum that distinguishes them from other exploding things, like exploding massive stars or exploding neutron stars. Most importantly, they are visible from across the universe. When one goes off in a distant galaxy, we can then measure with pretty good accuracy just how distant that galaxy is. We can also measure the speed at which that galaxy is moving away from us, and now we have a new point for our plot of speed versus distance.

Observations of Type Ia supernovae with the Hubble Space Telescope in the late 1990s and early 2000s found something unexpected. The most distant galaxies that we could get good distance measurements for were moving away from us faster than they should have been if the expansion of the universe were constant. That lovely straight line that Hubble made turns out to start curving upward for galaxies far from us. This means that the expansion of the universe isn't a constant but is changing in a way that indicates acceleration. We don't know what is driving the acceleration. We label it dark energy and hope that future observations will clarify the exact nature.

One question that we thus far haven't addressed is: what is the universe expanding into? What's kind of funny about this question, at once silly but profound, is that the answer seems to be both nothing and the universe. If the universe is infinite, as we believe, then there is no edge and no center and thus nothing to expand into. Potentially the word "into" is the problem here—if spacetime is expanding, then it's not expanding into anything; it's just getting bigger. If there is an edge of the universe, we may never observe or experience it, and thus the question is as unanswerable as asking someone standing at the North Pole which way north is. The question becomes almost a category error. It's closely related to the question of what happened before the Big Bang. These questions are, at least now, more in the realm of philosophy than physics. Unanswerable with our current models and understanding, but maybe one day open to us. A Nobel Prize awaits whoever can give us a real answer to either.

Finally, now we get to the heart of things. What will happen to our cozy and beautiful and brilliant universe, one that has indifferently and yet benignly fostered our own existence? The eventual fate of our universe is probably somewhere between the Big Rip and the coasting universe—not a universe that tears itself apart, but neither one that eventually reaches stasis.

The eventual fate of our universe is still one of complete loss and eventual motionlessness, a complete inability for anything to ever happen again. First, we have the period we are in now, dominated by the formation of stars and the existence of galaxies. This "Stelliferous Era" will likely last up to 100 trillion years, so we are only at the start of it. Over this era, the supply of available hydrogen will slowly dwindle, resulting in fewer and fewer stars being formed until the fuel is finally exhausted, and no new stars will ever form. As larger stars die off, only the smallest stars will burn on. Eventually, at the end of the age of stars, even those most efficient and longest-lived stars will die out.

Meanwhile, the Milky Way and the Andromeda galaxies will merge in about 4.5 billion years, forming a new galaxy currently dubbed "Milkdromeda," just as our Sun is ending its life. In 150 billion years, because of the accelerating expansion of the universe, all galaxies outside of the local supercluster of galaxies will no longer be observable. In about one trillion years, all of the galaxies in the "Local Group," a collection of nearby galaxies, will end up merged into one big Milky-Andromedy-Local galaxy. In about two trillion years, all galaxies outside of the Local Group merged galaxy will no longer be observable. What's left is up to ninety-eight trillion years of low-mass stars shining in the only galaxy that will be observable to whoever is left there to observe it.

After the last stars have winked out 100 trillion years after the beginning of time, all that remains is degenerate matter—white dwarfs, neutron stars, and black holes. All remnants of an early and brighter age. The universe will become dark, punctuated by only occasional flashes of intense brightness as white dwarfs or neutron stars happen to merge and exceed their mass limits, forming rare but luminous supernova explosions. In this dark, cold, and barren universe, the infinity of time eventually destroys all things. Planets end up flung away from their long-dead stars through gravitational interactions or their orbits slowly decay, and they crash into the cinder of their star. One quadrillion (or 1,000 trillion) years have passed since the Big Bang and all planetary systems are gone. Eventually, all objects left in the Local Group merged galaxy will suffer one of two fates—either their orbits decay and they merge into the black hole at the center of the merged galaxy, or they are also flung out of the galaxy by gravitational interactions. Any objects eaten by the black hole will briefly reignite the quasar, temporarily sending light out into an increasingly empty universe.

The next step depends on whether protons decay or not. If protons do not break down into smaller parts, the universe takes a long time to settle into its final state. All non-black hole matter will eventually become iron, through slow and random quantum tunneling. Iron stars will eventually become black holes, and all black holes will eventually evaporate away into nothing. In the Dark Era, with all matter converted into subatomic particles, the universe is an almost pure vacuum and has settled into a cold and static state. This may take as long as 10^10^120 years (that's ten to the power of ten to the power of 120, a number that even a brain that is used to large numbers can't comprehend). The universe stops changing and becomes both sad and boring.

If protons decay, the actual final state is nearly the same but on a shorter timeline. All atoms will eventually fall apart, converting into photons and subatomic particles. Black holes will remain, indifferent to proton decay, but still will eventually evaporate. 10^{100} years after the Big Bang, the universe will reach its end state of vacuum, subatomic particles, and extreme coldness. No activity will be possible, as the heat death of the universe is finally achieved. This heat death is a state of completely uniform distribution of energy, where there is no thermodynamic energy available to do any work. Heat death is unrecoverable.

What this means for us: In either case, our current period of light and stars and galaxies, the magnificent universe we can observe and try to understand, is over in the blink of an eye. The universe is only briefly this beautiful. It's hard to comprehend that this age of time, even if it can be measured in trillions of years, in hundreds of times more years than have even happened yet, is a fleeting instant in the lifetime of the universe. We will be longer than long gone when the last stars finally die, and no trace of our existence will ever be preserved against the laws of physics. Yet, I feel an incredible loneliness to contemplate this dismal and quiet end. All of this majesty and beauty and light only to inexorably end up in the cold and the dark. A sad end to the glory of the universe.

CHAPTER 15

The Dark Sector: The Weirdest Items on a Large but Incomplete List of All the Things We Don't Know

Weird Facts: Most of the universe is made up of matter and energy that we don't understand. We can only even tell that it exists at all because of the effect on things like stars and galaxies. There are two types: Dark matter is a mysterious form of mass that stubbornly refuses to interact with light or matter like what makes up stars and planets. Dark energy is even more mysterious and appears to be a strange property of spacetime that is causing the universe to expand faster and faster.

As we've just learned in the previous chapter, the fate of the universe is driven by the exact presence and nature of a mysterious energy that seems to pervade all of spacetime. We don't know what this "dark energy" is, if it's an intrinsic quality of spacetime, if it's a strange energy density that is variable in time and space, or if it's not real at all and just an observational effect that hasn't been correctly calibrated and accounted for.

Dark matter and dark energy are similarly named, but aside from their names, they describe unrelated effects—they have no overlap or similarity aside from both being mysterious to us. "Dark matter" was named earlier—first postulated and named in the 1930s and then measured in the 1960s. The term "dark energy" was a deliberate call back to dark matter, both unknown behaviors only observed at the largest scales with the most careful observations.

We've already had an introduction to dark energy, but let's be precise about a few additional things that we know. The first is that all we can see is the distant effect of dark energy on the farthest galaxies. From this, we must infer what is happening, but we have no direct measurements of dark energy itself. The possible physical explanations for it are variants of the same theme—that the energy is a property of space itself. Sometimes this is referred to as a "cosmological constant," after a coefficient term that Einstein used in some early versions of general relativity to ensure a static, unchanging universe. Once the universe was discovered to be expanding, Einstein removed the constant (technically he set it to zero) and referred to it as his "greatest blunder." Physically, the constant just provided a way to counterbalance the effect of gravity in the calculations of the behavior of spacetime. After the discovery of the accelerating universe, the constant returned, to provide this property of space. The constant is now referred to as Lambda, usually by the capital Greek letter .

Imagine a perfect cube of empty space; no particles inside, just an empty volume floating out somewhere far from anything. If you could put a little plexiglass box around it, space would push outward with a tiny bit of force. If you don't have a little plexiglass box, the effect is the same, but the volume of space will grow a tiny bit instead. Sometimes this is referred to as vacuum energy. In fact, quantum mechanics predicts that the vacuum should have some intrinsic energy, as particle-anti-particle pairs pop in and out of existence in a low-level quantum churn. However, the measured amount of vacuum energy from cosmology disagrees with any attempts to use quantum field theory to predict a value. Other variations of dark energy theory have the same effect (vacuum energy), but the reason is different. Quintessence is one such model, where rather than being an intrinsic property of space, quintessence is a field that can vary across space and time.

All of these theories about dark energy have the use of general relativity at their core. Our assumptions about the behavior of the universe at large scales all use general relativity and its equations for every calculation. Some of the more outlandish theories to explain dark energy are revisions of general relativity. In these cases, modifications to gravity would eliminate the need for dark energy at all (and some also eliminate dark matter), explaining the behavior of distant galaxies by changes in gravity alone. But, none of those attempts have been convincing or able to explain all observations. General relativity, despite its known problems, has stood through every possible test thus far.

Interestingly, we already know general relativity is not perfect: it fails at predicting the nature of spacetime at a singularity. In science, even one failure of a theory means that it needs to be changed and fixed. So, we already know there must be a more comprehensive theory of the universe. We just don't know what that theory is yet. General relativity was a refinement of Newton's Universal Theory of Gravity, so there is nothing at all which says we can't overthrow general relativity and develop something better. We haven't figured it out yet, but I think it's a matter of time. It took hundreds of years, extremely laborious data collection, and careful experimentation and observation to create a body of knowledge that set the scene for Einstein to develop general relativity. It likely will take just as much work and effort for the next thing, but it will happen.

Compared to dark energy, dark matter is simpler to understand and has much less of an impact on the overall fate of the universe. Dark matter also has a slightly better connection to its name. Dark matter is some type (still TBD) of matter that does not interact with light, thus we can't see it the way we can see normal matter. Normal matter (in the world of astrophysics, we refer to normal matter as baryons) consists of protons and neutrons, the subatomic particles that, along with electrons, form the atoms and molecules that make up your body, this book you're reading, the chair you are sitting in. Baryonic matter can interact easily with light—if you shine a light on normal matter, some light gets absorbed, some gets reflected, some gets transmitted. The path of the photon is shifted in some way by the interaction, and we can observe that effect. Baryons also interact with themselves—two particles will fuse or bond or bounce off each other like billiard balls. This interaction is a key reason we can observe them—typically these interactions give off small amounts of energy or photons and we can observe those photons as a record of the interaction.

Dark matter does not appear to interact with photons, it doesn't interact with normal baryonic matter, and it may not even interact with itself. But it does appear to follow the rules of gravity—dark matter interacts only gravitationally. We know this because that's the only way we can observe it at all—via its gravitational effect on normal matter.

Dark matter was first discovered and observed in the 1960s by American astronomer Dr. Vera Rubin. Rubin was a trailblazer for women in astronomy, becoming the first woman to observe at historic telescopes such as the Palomar Observatory, at the time the most powerful telescope on Earth. A tragic and true story, but the dorm at Palomar, on a mountaintop in Southern California, was called the "monastery" because historically only men observed there. It's still called that today even though people of all genders observe and work there.

Rubin was observing the motion of stars at the outskirts of nearby galaxies. She started by observing Andromeda, the nearest large spiral to our Milky Way. She found that the motion of stars far from the center of the galaxy was faster than expected. We've known for hundreds of years that you can measure the mass of one object by observing another object orbiting around it. The orbit is dependent on just gravity and the mass of the central object. We can do this mass measurement for the Sun using the orbits of planets. We can also do this measurement for galaxies by using the orbits of stars. If you use a star close to the center of a galaxy, you'll only get the mass of the galaxy within the orbit of the star. If you use a star at the edge of the galaxy, you should be able to measure the mass of the whole galaxy. Rubin observed that using the motions of stars around their home galaxies, every galaxy was more massive than previously understood. These galaxies were five, sometimes ten, times as massive as they should be given the number of stars within the galaxy. Because of this extra mass, the galaxies were rotating faster than they should have been. They were spinning so quickly that without this mysterious extra mass, they would have spun apart. Rubin found that every galaxy she observed had this extra mass.

Rubin had found evidence for dark matter. The effect of dark matter has been seen in its gravitational impact at other scales—in clusters of galaxies, in merging galaxies, even in galaxies at the edge of the universe. In general relativity, mass bends spacetime and this can create the effect of a lens, magnifying background objects as the light passes nearby to a mass in the foreground. Using this gravitational lensing, we can measure the mass of galaxies and clusters and still find that there is too much mass compared to just the mass of stars. We also see it in the large-scale structure of the universe—the way that galaxies form and condense on large scales happened because of the noninteracting nature of dark matter in the early days of the universe when it was too hot and dense for normal matter to condense.

Counts of black holes, planets, and other objects that are made up of normal matter but are not in stars just don't yield nearly enough material to account for the mass of dark matter. Aside from the gravitational effects, there is no other way to measure dark matter. That's why we describe it as noninteracting. If a particle of dark matter flew through your body right at this moment, you wouldn't feel it. It's a ghost to you. On large scales, the mass makes a difference, but on the scales of the solar system or even the nearby galaxy, the effect is so tiny as to be unnoticeable. Rubin never won a Nobel Prize for this work, despite being one of the more deserving people of it in the last sixty years.

The exact nature of dark matter is unknown. We think it's a particle, but one that hasn't been found yet by particle accelerators or the standard model of particle physics. Theories abound (there are so many of them)—including amusingly named ones like MACHOs (massive compact halo objects), WIMPs (weakly interacting massive particles), fuzzy cold dark matter, sterile neutrinos, and more. It's a rogue's gallery of theories, each one difficult to rule out yet difficult to prove. Other versions of the explanation again go back to modifications to how gravity behaves on large scales. These modifications to gravity have not yet addressed things like the mass measured through gravitational lensing and still don't give a compelling picture.

Combining the effect of dark matter and dark energy, the "dark sector" of the universe occupies an overwhelming fraction of the universe's energy density. This quantity is a measurement of how much energy there is per volume in the universe, averaged over large scales. Since $E=mc^2$ tells us that mass and energy are the same, we can convert all the mass (dark and not dark) into energy and figure out the total energy budget of the universe. For our entire observable universe, dark energy is 68.3 percent of all energy, while dark matter is 26.8 percent. The remaining 5 percent of the energy balance is the normal, baryonic matter that has occupied nearly this entire book. That's it. The entirety of your existence and all that you can see, observe, and experience is a tiny, relatively insignificant fraction of what's in the universe. It also means we have almost no idea what makes up 95 percent of the energy content of the universe. In terms of things we don't know in the universe, dark matter and dark energy may be only two items on a list of hundreds, but they weigh a lot.

Both dark energy and dark matter are fundamentally signatures of nature's weirdest force, gravity. The fact that we don't understand gravity at all is probably why dark matter and dark energy give us such difficulties. Gravity is by far the wimpiest out of the four fundamental forces. It's so weak that you can overcome the *entire* gravitational effect of the Earth by just jumping. That's a weak force. In contrast, strong magnetic or electric fields will just kill you if they are strong enough (we've seen this in the magnetic fields found in magnetic neutron stars).

But despite being weak, gravity is the shaping and meaningful force in the long run because it can act over long distances. Technically gravity can act over an infinite distance—the force becoming smaller and smaller with distance but never truly zero. This long-range interaction is why gravity matters so much in space and why it was the first force that we noticed and observed. General relativity gives us the most comprehensive version of gravity, not as a force (the way that the electromagnetic force is a force), but as an effect of the curvature of spacetime that mass creates. Why does mass curve spacetime? Because we live in a spacetime where mass or energy can cause curvature. Why does that happen? Well, we don't know the answer to that yet. Maybe that will be answered by quantum gravity one day. Or string theory. Or by other theories that predict many dimensions and gravity, which could be incredibly strong in some other dimension. For now, we are left with this wonderful and weird force that is incredibly weak but can still reach across the entire universe, one that has driven the creation of stars, planets, galaxies, and black holes. One that is part of the reason I exist to write this book and you exist to read it.

What this means for us: We've returned to this time and again in this book, but fundamentally our understanding of the universe is an understanding of gravity and its many facets. Gravity in tiny and dense spaces, gravity in a star, gravity in a galaxy, gravity across the universe. Most of the unknowns in our understanding (including dark matter and dark energy) are related to our understanding of gravity. When we one day have a complete and full understanding of the nature of gravity, only then we will truly know the universe.

CHAPTER 16

We Are the Universe Observing Itself

Dark energy, this mysterious and seemingly pervasive quantity, is one item on a long list of things we don't understand about the universe. Its partner in crime, dark matter, is, of course, high on that list. But there are so many more things we don't know about the universe we live in. Some of these unknowns are about the future—will it ever be possible to travel between the stars? Some of these unknowns are about the state of spacetime—why is the speed of light the universal speed limit? What happens at the singularity of a black hole? Why is spacetime flexible at all? Some of these unknowns are about our origins—how common is life in the universe? Are we alone? And some of these unknowns are the most basic questions—what is gravity? How did the universe start? Why is there anything at all in this universe? Is there an edge to the universe? Is our universe the only one?

It's a mind-bending list of questions. Some of them are seemingly so simple, yet they reveal how little we know about anything. We have touched on bits and pieces of these throughout this book, but if one question is causing your brain to itch a little, I encourage you to explore more beyond these short pages. The discoveries we have made in our understanding of the universe have all been made by people like you—people who are curious and maybe a little annoyed about our lack of understanding.

If we consider the state of our understanding of the universe from 100 years ago—1925—we can see how far we've come. We've learned what powers the stars; we've discovered degenerate states of matter and found neutron stars and white dwarfs out in the galaxy; we've found actual black holes; we've discovered that planets are stunningly common; we've learned that the universe had a starting point; that it's filled with weird types of matter and energy that we don't understand; we've seen and measured ripples and waves in spacetime; we've discovered a slew of subatomic particles and peered into the hearts of atoms. The level of advancement even in the last thirty years is incredible. The signatures of dark energy were only discovered in the late 1990s. We measured the first ripples in spacetime caused by two black holes merging less than ten years ago. We took the first picture of an accretion disk around a supermassive black hole just five years ago. I write these because it can be so easy to forget how much progress science makes every day. The breakthrough advances are easy to see in hindsight. How thrilling to imagine what we will know 100 years from now.

In humanity's attempts to understand the universe around us, we are trapped in an interesting paradox. I hinted at this chapters ago when describing just how important certain properties and constants are. Why do certain constants have the values that we observe them to have? Change the composition of the early universe a little and galaxies never form. Make the strong force (which keeps atomic nuclei together) a tiny bit stronger and the early universe would have converted all hydrogen into helium during the first twenty minutes, leaving no hydrogen to form the stars we know today or to make water on Earth. In some sense, our universe feels finely tuned to support the formation of stars and planets and maybe life. This fine-tuning is referred to as the anthropic principle. The anthropic principle argues that life, or even inanimate objects like stars and planets, will never form in a universe that doesn't have the right conditions for that formation, and so, just by being here and being able to observe our universe, we are already biased in what we observe. A human will never evolve in a universe that can't form stars; thus, a human will never observe or measure a universe that can't form stars. That doesn't mean those universes don't exist (or couldn't exist in some other part of our universe). It just means we will never observe them because it excludes us.

The anthropic principle is not a real scientific statement—it can't be falsified or proven wrong. It's more of a way of thinking about the current limits of our observations. A more succinct description is that the entire universe suffers from a selection effect. We are only observing it because the universe contains conditions for us to do so. The title of this chapter is an echo of this and a rephrasing of the quote by Alan Watts: "Through our eyes, the universe is perceiving itself. Through our ears, the universe is listening to its harmonies. We are the witnesses through which the universe becomes conscious of its glory, of its magnificence."

Frequently people ask me, why study the stars? Why explore space? Why wonder about the nature of dark matter and energy? Why spend money to send telescopes and satellites and even people out into the void? What I frequently say is that there are many reasons, some more practical than others. The most practical reason is that basic research like astronomy yields discoveries and inventions that can't be known ahead of time. My favorite example of this is the link between the Global Positioning System (GPS) and general relativity. Einstein developed general relativity because he wanted to explain how the universe worked more accurately. He didn't do it for any commercial reasons. Yet our ability to use GPS now, 100 years later, is reliant on the calculations of general relativity. Satellites in orbit experience a slightly different spacetime than you, on the surface of the Earth. Those differences, on the order of nanoseconds, matter for GPS calculations. Without general relativity, GPS would be off by miles. So first, who can say that discoveries that seem esoteric now won't one day be incorporated into the infrastructure of our lives?

But beyond this, I don't believe science or discovery needs to be utilitarian to be important and valuable. I believe that exploring the universe, asking questions about it, wondering about it, is as fundamental a part of the human experience as love or desire. For all of human history, except for the last roughly 150 years of inexpensive electricity, every human experienced the full majesty of the night sky nearly every night. Our ancestors evolved seeing the stars, the Milky Way, this vast carpet of points in the darkness. They were so preoccupied with these points of light that they tracked them carefully, noticing which ones fit patterns and which ones didn't. They worshipped them. Every culture on Earth independently made up and passed down stories about these points of light; they all gave them names. They used them to figure out when to plant, move, and expect different weather. The ability to predict and understand the stars was so important that astronomy is one of the oldest professions in the world.

Looking up at the night sky and wondering is one of the most intensely human things that one can do. What is more human than attempting to wrestle the vastness of the universe into understandable rules? What is more human than exploring the limits of what light can probe? What is more human than doing so for thousands of years, with the expectation that just this time, we will figure it out? Hubris, intense curiosity, pattern recognition, self-importance, our best and worst qualities are all part of the motivation.

I believe that astronomy, like food and shelter and security, is a requirement for human life. We observe because we must. That's why we study the stars. I hope too that one day we will discover ways to explore and expand our understanding far beyond its current limits. But even if that day doesn't come for me, I am satisfied to know just a little more about how the universe works. I look at images of stars, of nebula, of other galaxies, I see the night sky from my backyard in Arizona, and I see how shockingly beautiful the universe is. I feel grateful that I can experience it. I think about how lucky we are to be here to experience it at all, how unlikely every step of the way is, and how glad I am to have this short time here on Earth to see it.

"We are a way for the universe to know itself. Some part of our being knows this is where we came from. We long to return. And we can, because the cosmos is also within us."

—CARL SAGAN

CONCLUSION

The Earth Isn't at the Center of Anything

The universe is weird and wonderful. We are so lucky to exist in it, to be able to observe and experience just the tiniest fraction of its beauty and majesty. While we've discovered so much about the state of the universe, the planets and stars and galaxies that it contains, there is so much more we still need to find out. As evidenced by the many twists and turns in the history of science, I hope and expect that discoveries over the next 100 years will continue to attest that the universe is weird and unpredictable at its very core. Some possible discoveries to look forward to: Will we ever figure out the nature and behavior of gravity and why it's so weird? Will the behavior of singularities be puzzled out? Will we explain dark matter? What is dark energy?

All these questions are important and exciting, but they are also the tip of the iceberg. Even more interesting to me are the questions that we don't yet know to ask. What will be discovered in the future that sets off the next set of new questions?

If you want to learn more about the weirdness of the universe, there are so many places to turn: classics like the TV show *Cosmos*, hosted by Carl Sagan; modern videos like those from Kurzgesagt and PBS Spacetime; books by Neil DeGrasse Tyson, Janna Levin, and many more. The most important thing is to be curious, spend time learning and thinking, and keep asking questions.

Lastly, the best place to experience the wonder of space is outside, under the stars themselves. Even if you live in a city, the stars are there. Take a minute tonight to go outside, find a dark place, and look up.

About the Author

Bret Hartman / TED

Dr. Erika Hamden is a professor of astrophysics at the University of Arizona and director of the University of Arizona Space Institute. She builds telescopes, mostly for NASA, to observe the universe and understand how stars and galaxies form. Professor Hamden is a world expert in space sciences, including space mission development and ultraviolet technology. She earned a BA from Harvard College and a PhD from Columbia University. She is a TED Fellow, an AAAS IF/Then Ambassador, and has won numerous awards from NASA for her work in ultraviolet astronomy. In addition to her technical work, she hosts a TV show on AZPM, holds a private pilot's license, was previously a professional chef, and teaches science on social media to over 170,000 followers. She lives in Tucson, Arizona.

Mango Publishing, established in 2014, publishes an eclectic list of books by diverse authors—both new and established voices—on topics ranging from business, personal growth, women's empowerment, LGBTQ studies, health, and spirituality to history, popular culture, time management, decluttering, lifestyle, mental wellness, aging, and sustainable living. We were named 2019 *and* 2020's #1 fastest growing independent publisher by *Publishers Weekly*. Our success is driven by our main goal, which is to publish high-quality books that will entertain readers as well as make a positive difference in their lives.

Our readers are our most important resource; we value your input, suggestions, and ideas. We'd love to hear from you—after all, we are publishing books for you!

Please stay in touch with us and follow us at:

Facebook: Mango Publishing
Twitter: @MangoPublishing
Instagram: @MangoPublishing
LinkedIn: Mango Publishing
Pinterest: Mango Publishing
Newsletter: mangopublishinggroup.com/newsletter

Join us on Mango's journey to reinvent publishing, one book at a time.